Nina Zhang

Revised Printing

ENVIRONMENTAL SCIENCE
Laboratory Manual

Tatyana A. Lobova
Old Dominion University

Kendall Hunt
publishing company

TA: atummoo2@odu.edu

Cover image © Shutterstock, Inc. Used under license.

Kendall Hunt
publishing company

www.kendallhunt.com
Send all inquiries to:
4050 Westmark Drive
Dubuque, IA 52004-1840

Copyright © 2011 by Kendall Hunt Publishing Company
Revised Printing 2014.

ISBN 978-1-4652-5295-1

Printed in the United States of America
10 9 8 7 6 5 4 3 2 1

CONTENTS

ABOUT THIS LAB MANUAL AND NOTEBOOK

This Lab Manual and Notebook was specifically created for the laboratory portion of the course BIOL 111N *Environmental Science* at Old Dominion University (ODU). The Lab Manual provides students with the information necessary to conduct activities and hands-on experiments that will enhance their understanding of the major concepts and issues of Environmental Science. The content introduces students to the basics of scientific inquiry, scientific method, and hypothesis testing—the universal approaches that scientists use to address the questions about the empirical world. In addition, this volume also serves as a lab Notebook.

This Lab Manual consists of three main sections: course introductory information, labs, and appendices.

1. Course introductory materials include standard Lab Safety Rules, which are universal for all science labs at Old Dominion University, Lab Policies that are specific for BIOL 111N lab, and Emergency Information that is kept in the office of the Department of Biological Sciences.

2. Labs 1 through 10 consist of the activities related to one or more Environmental Science topics. Each lab is generally organized following this format:

    Pre-lab Reading: Selection of topics from the textbook and material from the Lab Manual that need to be read prior to the lab in order to better understand and complete lab activities.

    Activities: List of activities included in the current lab.

    Introduction: Brief background information and major terminology on the topic of the current lab; also a purpose and an overview of the exercises.

    References: Sources that were used for the preparation of this lab topic.

Objectives: List of what students are to learn as a result of completing the specific activity.

Hypotheses: Hypothesis or hypotheses that are being tested in this activity.

Materials: List of materials that are necessary to conduct the activity.

Procedure: Step-by-step description of the activity.

Data: Pages provided to record the raw data and observations for each activity.

Discussion and Conclusion: List of questions pertaining to the completed lab activities that require students to identify potential errors, interpret the collected data, compare the individual data to the group or class data, and draw conclusions related to the original hypotheses. These pages should be completed by students **individually**, removed from the Lab Manual, and submitted to the TA at the end of the lab for completion points.

3. Appendices include information on group projects and blank data sheets.

LAB SAFETY RULES

The following policies will be strictly enforced.

- Become thoroughly acquainted with the location and use of safety facilities. Emergency eyewash stations are located in each laboratory, as an attachment to a faucet on a major sink in each laboratory. Each lab is equipped with a safety shower or has a safety sticker denoting the nearest shower. Use these showers only in emergency and clear the area (lab) when using, as there are currently no floor drains and the floor will be flooded in a matter of minutes. Each laboratory is equipped with a fire extinguisher, which is inspected regularly by the Office of Environmental Health and Safety.
- In the event the **fire alarm** sounds, turn off any burners or hot plates that are in operation and proceed in an orderly manner through the nearest exit and evacuate the building. Do not use the elevator. Do not stand near the loading dock, as this is the triage area and the area where the fire engine will first respond.
- **NO sandals, flip-flops, or open-toed shoes; NO shorts;** and **absolutely NO** food, drink, or smoking allowed in the laboratory. Be prepared with an old shirt, a lab coat, a pair of sweat pants, and/or sneakers if your dress for the day is inappropriate for a lab setting. Long hair must be tied back. Students with tank tops, bare midriffs, or skimpy shorts will not be admitted to the lab. That means you will be unable to take the quiz or participate in the lab—so you will lose points!
- **All cell phones must be turned off** before entering the lab! Students using cell phones during the lab or leaving the phone on the bench top will be penalized.
- In the event of an accident, the laboratory instructor should be informed immediately. Minor first aid treatment is available at the B.S.S.F. room 207. Please note that the ODU Student Health Center will accept any student with or without insurance coverage on an emergency basis.

- In case of skin contact with an acid or a base, **WASH IMMEDIATELY** for at least 15 minutes. It may seem to be an excessive amount of time but is absolutely necessary to prevent severe blistering and/or burns. Chemicals spilled on clothing should be dealt with appropriately to prevent contact with skin.

- Before beginning an experiment, become familiar with the method of operations and all potential hazards involved, including biohazard level, flammability, reactivity, toxicity, and corrosiveness of material. Also be familiar with chemical and biological waste procedures.

- **Notify your instructor immediately** if you are pregnant, color blind, allergic to any chemicals, or have any medical condition (such as diabetes, immunologic defect, etc.) that may require special precautionary measures in the laboratory.

- Upon entering the lab, place all coats, purses, backpacks, etc. at the back of the room and hang on the hangers when possible. Placing of these personal items on the bench top or under the bench is not permitted.

- Report all spills and accidents to your instructor immediately.

- Never leave heat sources unattended.

- All students at a lab table will lose credit for the entire lab if they fail to clean up their lab station before the last student leaves the class.

- All students at a lab table will lose credit for an entire lab, as well as bear the cost of replacement, if any equipment is found missing from that table at the end of class.

- Do not perform experimental work in the laboratory alone.

- Do not perform unauthorized experiments.

- Do not use equipment without instructions.

- All honor code violations (or suspicions thereof) will be reported and pursued to the full extent provided by University policy.

Additional laboratory policies are included below. All students are responsible for reading, understanding, and following those policies.

LAB POLICIES

LATE ARRIVAL IN LAB

Students arriving late to laboratory disrupt the TAs and other students in the lab. Students are individually responsible for ensuring that they take whatever steps are necessary to be on time (this includes taking traffic and parking into account). Late arrivals will be dealt with as follows:

▶ When a Quiz is Scheduled

1. Quizzes will be handed out no more than 5 minutes after the "official" beginning of lab.
2. Students who arrive before the quiz is handed out are considered to be on time and are not penalized.
3. Students who arrive after the quiz is handed out but before the lab lecture begins must remain outside the lab and will NOT be allowed to take the quiz; they will NOT be allowed to make it up. They will be admitted to the lab after the quiz is over and will not otherwise be penalized.
4. Students who arrive after the quiz is over and the lab lecture has begun will NOT BE ADMITTED to the lab. They will not be allowed to turn in assignments (and will receive 0s). They will still be held responsible for the material covered on the quiz (if any) scheduled for the following week.
5. Graduate TAs may modify these guidelines at their discretion under ***extraordinary*** circumstances (i.e., extraordinary traffic delays vs. the usual slow-downs that happen during bad weather, last-minute family illness or other emergency, etc.).

▶ When a Quiz Is Not Scheduled

1. Students arriving more than 10 minutes after the start of lab will NOT BE ADMITTED to the lab; they are considered absent without excuse.

They will not be allowed to turn in their assignments (if any). They will still be held responsible for the material covered on the quiz (if any) scheduled for the following week.

2. Graduate TAs may modify this guideline at their discretion under ***extraordinary*** circumstances (i.e., extraordinary traffic delays vs. the usual slow-downs that happen during bad weather, last-minute family illness or other emergency. etc.).

STUDENT DISRUPTION DURING LAB

The following policy is based on the Old Dominion University Code of Student Conduct:

1. A student who talks, laughs loudly, or who otherwise disrupts a lecture or other lab activity, will receive one clear verbal warning that his/her behavior is disruptive and must stop.
2. If the student persists in those activities, she/he will be penalized 1–5 points depending on the severity of the disruption.
3. If the student still persists in disruptive activity, she/he will be asked to leave the classroom and will receive a score of 0 for all lab activities. If she/he refuses to leave, the TA will contact campus security to have the student escorted from the room. The student will be written up with an honor code violation in accordance with ODU policy.

MISSED LABS

To accommodate students with legitimate reasons for missing a lab, we drop your lowest quiz and completion-points score prior to calculating your final lab score. As with lecture exams, the first item you miss is the one we drop. Thus, even if you miss a lab for a legitimate reason, we will not give makeup quizzes or adjust scores. If you miss more than one lab for legitimate reasons, we will make arrangements on a case-by-case basis. Please review the following policies carefully and contact your TA if you have questions.

If you miss a lab, the only way you can make it up is to arrange to attend another lab later in the **same week**; we have no mechanism to allow you to make up an activity after the week in which it is normally conducted. Labs may be made up only if you miss due to a legitimate, documented reason. YOU MUST CONTACT YOUR TA WITHIN 24 HOURS OF YOUR REGULARLY SCHEDULED LAB to make the necessary arrangements unless you can demonstrate that the nature of your emergency prevents you from doing so. If you fail to contact your TA within 24 hours of missing your lab, you will not be allowed to make up the lab.

You should first attempt to schedule your makeup with your own TA. If you cannot attend his/her lab, you may contact another TA and see if you can attend another lab. You may only attend another lab with the PRIOR PERMISSION of BOTH your own and the other TA. We will work with students with unusual circumstances on a case-by-case basis.

1. If you miss a lab for a legitimate, documented reason and CANNOT make it up, you
   a. may not make up the quiz (the first quiz missed will be dropped);
   b. may not make up the lab completion question;
   c. may turn in assignments due that day, but only as specified via special arrangement with your TA. You are responsible for contacting your TA within 24 hours of the lab and following his/her instructions precisely.
   d. may take the quiz that is given the following week in your regular lab.

2. If you miss lab for a legitimate, documented reason and make it up with another TA (with that TA's prior permission!) you
   a. still need to turn in that week's assignment (if any) to your regular TA in a timely manner (your TA will provide information on how to do this when you notify him/her that you've missed lab);
   b. may take the OTHER TA's quiz during the lab you attend; the other TA will provide your TA with the relevant grades for the day;
   c. may take the OTHER TA's lab completion question;
   d. may take the following week's quiz with your TA in your usual lab.
3. If you miss a lab for a nonlegitimate or undocumented reason and/or fail to notify your TA within 24 hours of missing a lab, you will lose the quiz points for that lab, will be unable to turn in (or receive points for) assignments due that day, and will lose completion points.

LAB PENALTIES

Point penalties will be assigned as indicated for each occurrence of the following behaviors:

1. Late submission of written assignment: 5 points per day
2. Failure to bring a Lab Manual: 5 points
3. Disruptive behavior and/or failure to participate fully in lab activities: 1–5 points
4. Use of the cell phone while in the lab: 5 points
5. **Inappropriate use of lab computers: 10 points for each student in the lab group**

HONOR CODE STATEMENT

The following statement should be typed verbatim on all written work (quizzes, projects, etc.) and signed:

In completing this assignment I have abided by the Honor Code of Old Dominion University.

Lab 1
SCIENTIFIC INQUIRY

PRE-LAB READING

Textbook: Chapter 1: The Nature of Science.

Lab Manual:

- Lab Safety Rules and Other Policies.
- Lab 1. You will need to answer the questions in Activity 1 and Activity 2 to the best of your knowledge after the reading. Write the answers under the questions directly in the Manual. The questions and answers will be discussed in the lab.

ACTIVITIES

Lab Safety Rules and Course Overview & Policies

1. Defining a Problem
2. The Elements of an Experiment
3. Designing an Experiment
4. Critique of the Sample Proposal

INTRODUCTION

Scientific inquiry is a particular approach to answering questions about empirical world, that is the world that can be measured and observed through our senses (or instruments that extend them). Therefore not all the questions that may come to our mind in everyday life can be answered using scientific inquiry. Scientific inquiry starts with observations and questions and involves specific types of reasoning and evidence. If a scientific investigation is designed according to a traditionally accepted set of rules and conducted following the main steps of scientific method, then the results of this experiment can be tested and accepted by other scientists. If an investigator fails to follow the guidelines or would not disclose all the details and elements of the scientific protocol, then the results of this experiment will not be recognized as valid by the scientific community.

In this laboratory you will be introduced to the safety rules of the Biology lab and will be given an overview of the course structure and policies. During this lab you will learn about the basic elements of scientific inquiry and how to apply this process to answering questions.

REFERENCES

Dickey, J. 2003. *Laboratory Investigations for Biology*. 2nd edition. Benjamin Cummings, San Francisco, California.

Lab 1

ACTIVITY 1
Defining a Problem   Nina Zhang

OBJECTIVES

After completing this exercise, you should be able to

1. Identify questions that can be answered through scientific inquiry and explain what characterizes a good question.
2. Identify usable hypotheses and explain what characterizes a good scientific hypothesis.

Every scientific investigation begins with the question that the scientist wants to answer. The questions addressed by scientific inquiry are based on observations or on information gained through previous research, or on a combination of both. Just because a question can be answered doesn't mean that it can be answered *scientifically*. Discuss the following questions with your lab team and decide which of them you think can be answered by scientific inquiry.

What is in a sealed box? Y
Are serial killers evil by nature? N
What is the cause of AIDS? Y
Why is the grass green? Y
What is the best recipe for chocolate chip cookies? N
When will the Big Earthquake hit San Francisco? Y
How can the maximum yield be obtained from a peanut field? Y
Does watching television cause children to have shorter attention spans? Y
How did you decide what questions can be answered scientifically?
   Based on if we can make an observation for it or fact vs opinion.
   Is it measurable? can you test it?

A scientific question is usually phrased more formally as a **hypothesis**, which is simply a statement of the scientist's educated guess at the answer to the question.

A hypothesis is usable only if the question can be answered "no." If it can be answered "no," then the hypothesis can be proven false. The nature of science is such that we can prove a hypothesis false by presenting evidence from an investigation that does not support the hypothesis. But we cannot prove a hypothesis true. We can only support the hypothesis with evidence from *this particular investigation*. For example, you used hypotheses to investigate the contents of your sealed box. A reasonable hypothesis might have been, "The sealed box contains a penny and a thumbtack." This hypothesis could be proven false by doing an experiment: putting a penny and a thumbtack in a similar box and comparing the rattle it makes to the rattle of the sealed box. If the objects in the experimental box do not sound like the ones in the sealed box, then the hypothesis is proven false by the results of the experiment, and you would move on to a new hypothesis. However, if the two boxes do sound alike, then this does not prove that the sealed box actually

Dickey, Jean, *Laboratory Investigations for Biology, 2nd ed.*, © 2003, pp. 1.4–1.14, A13–A24. Reprinted by permission of Pearson Education, Inc., Upper Saddle River, NJ.

contains a penny and a thumbtack. Rather, this investigation has supplied a piece of evidence in support of the hypothesis.

You could test almost any hypothesis you made by putting objects in the empty box. What one hypothesis could not be proven false by experimentation?

The scientific method applies only to hypotheses that can be proven false through experimentation (or through observation and comparison, a different means of hypothesis testing). It is essential to understand this in order to understand what is and is not possible to learn through science. Consider, for example, this hypothesis: More people behave immorally when there is a full moon than at any other time of the month. The phase of the moon is certainly a well-defined and measurable factor, but morality is not scientifically measurable. Thus there is no experiment that can be performed to test the hypothesis. Propose a testable hypothesis for human behavior during a full moon.

Which of the following would be useful as scientific hypotheses? Give the reason for each answer.

Plants absorb water through their leaves as well as through their roots.

yes, because we can test how plants can grow.

Mice require calcium for developing strong bones.

yes, because it can be tested.

Dogs are happy when you feed them steak. *No, it's opinion based by dogs.*
An active volcano can be prevented from erupting by throwing a virgin into it during each full moon.

No, stasically, has that ever been tested? Is it safe for humans?

The higher the intelligence of an animal, the more easily it can be domesticated.

yes, different animals are smarter than others.

The earth was created by an all-powerful being.

No, who would be the all-powerful being? It's not testable.

HIV (human immunodeficiency virus) can be transmitted by cat fleas.

yes, it can be tested. HIV can be transmitted by animals I'm sure.

Lab 1

ACTIVITY 2 The Elements of an Experiment

Nina Zhang

OBJECTIVES

After completing this exercise, you should be able to

1. Define and give examples of dependent, independent, and standardized variables.
2. Identify the variables in an experiment.
3. Explain what control treatments are and why they are used.
4. Explain what replication is and why it is important.

Once a question or hypothesis has been formed, the scientist turns his attention to answering the question (that is, testing the hypothesis) through experimentation. A crucial step in designing an experiment is identifying the variables involved. **Variables** are things that may be expected to change during the course of the experiment. The investigator deliberately changes the **independent variable**. He measures the **dependent variable** to learn the effect of changing the independent variable. To eliminate the effect of anything else that might influence the dependent variable, the investigator tries to keep **standardized variables** constant.

independent – what is being changed
dependent – what is being measured

DEPENDENT VARIABLES

The **dependent variable** is what the investigator measures (or counts or records). It is what the investigator thinks will vary during the experiment. For example, she may want to study peanut growth. One possible dependent variable is the height of the peanut plants. Name some other aspects of peanut growth that can be measured.

– The ammount of peanuts growing in one area/space.

All of these aspects of peanut growth can be measured and can be used as dependent variables in an experiment. There are different dependent variables possible for any experiment. The investigator can choose the one she thinks is most important, or she can choose to measure more than one dependent variable.

Dickey, Jean, *Laboratory Investigations for Biology, 2nd ed.*, © 2003, pp. 1.4–1.14, A13–A24. Reprinted by permission of Pearson Education, Inc., Upper Saddle River, NJ.

INDEPENDENT VARIABLES

The **independent variable** is what the investigator deliberately varies during the experiment. It is chosen because the investigator thinks it will affect the dependent variable. Name some factors that might affect the number of peanuts produced by peanut plants.

In many cases, the investigator does not manipulate the independent variable directly. He collects data and uses the data to evaluate the hypothesis, rather than doing a direct experiment. For example, the hypothesis that more crimes are committed during a full moon can be tested scientifically. The number of crimes committed is the dependent variable and can be measured from police reports. The phase of the moon is the independent variable. The investigator cannot deliberately change the phase of the moon, but can collect data during any phase he chooses.

Although many hypotheses about biological phenomena cannot be tested by direct manipulation of the independent variable, they can be evaluated scientifically by collecting data that could prove the hypothesis false. This is an important method in the study of evolution, where the investigator is attempting to test hypotheses about events of the past.

The investigator can measure as many dependent variables as she thinks are important indicators of peanut growth. By contrast, she must choose only one independent variable to investigate in an experiment. For example, if the scientist wants to investigate the effect that the amount of fertilizer has on peanut growth, she will use different amounts of fertilizer on different plants; the independent variable is the amount of fertilizer. Why is the scientist limited to one independent variable per experiment?

Time is frequently used as the independent variable. The investigator hypothesizes that the dependent variable will change over the course of time. For example, she may want to study the diversity of soil bacteria found during different months of the year. However, the units of time used may be anywhere from seconds to years, depending upon the system being studied.

Identify the dependent and independent variables in the following examples (circle the dependent variable and underline the independent variable):

1. Height of bean plants is recorded daily for 2 weeks.

    DV = height
    IV = TIME

2. Guinea pigs are kept at different temperatures for 6 weeks. Percent weight gain is recorded.

    DV = % weight gain
    IV = temp.

3. The diversity of algal species is calculated for a coastal area before and after an oil spill.

DV = Diversity
IV = Before & after oil spill

4. Light absorption by a pigment is measured for red, blue, green, and yellow light.

IV = color
DV = Light absorption

5. Batches of seeds are soaked in salt solutions of different concentrations, and germination is counted for each batch.

DV = germination
IV = different concentration

6. An investigator hypothesizes that the adult weight of a dog is higher when it has fewer littermates.

IV = littermates
DV = weight of dog

STANDARDIZED VARIABLES

A third type of variable is the **standardized variable**. Standardized variables are factors that are kept equal in all treatments, so that any changes in the dependent variable can be attributed to the changes the investigator made in the independent variable.

Since the investigator's purpose is to study the effect of one particular independent variable, she must try to eliminate the possibility that other variables are influencing the outcome. This is accomplished by keeping the other variables at constant levels, in other words, by *standardizing* these variables. For example, if the scientist has chosen the amount of fertilizer as the independent variable, she wants to be sure that there are no differences in the type of fertilizer used. She would use the same formulation and same brand of fertilizer throughout the experiment. What other variables would have to be standardized in this experiment?

PREDICTIONS

A hypothesis is a formal, testable statement. The investigator devises an experiment or collects data that could prove the hypothesis false. He should also think through the possible outcomes of the experiment and make **predictions** about the effect of the independent variable on the dependent variable in each situation. This thought process will help him interpret his results. It is useful to think of a prediction as an if/then statement: *If* the hypothesis is supported, *then* the results will be . . .

For example, a scientist has made the following hypothesis: Increasing the amount of fertilizer applied will increase the number of peanuts produced. He has designed an experiment in which different amounts of fertilizer are added to plots of land and the number of peanuts yielded per plot is measured.

What results would be predicted if the hypothesis is supported? (State how the dependent variable will change in relation to the independent variable.)

What results would be predicted if the hypothesis is proven false?

LEVELS OF TREATMENT

Once the investigator has decided what the independent variable for an experiment should be, he must also determine how to change or vary the independent variable. The values set for the independent variable are called the **levels of treatment**. For example, an experiment measuring the effect of fertilizer on peanut yield has five treatments. In each treatment, peanuts are grown on a 100-m² plot of ground, and a different amount of fertilizer is applied to each plot. The levels of treatment in this experiment are set as 200 g, 400 g, 600 g, 800 g, and 1000 g fertilizer/100 m².

The investigator's judgment in setting levels of treatment is usually based on prior knowledge of the system. For example, if the purpose of the experiment is to investigate the effect of temperature on weight gain in guinea pigs, the scientist should have enough knowledge of guinea pigs to use appropriate temperatures. Subjecting the animals to extremely high or low temperatures can kill them and no useful data would be obtained. Likewise, the scientist attempting to determine how much fertilizer to apply to peanut fields needs to know something about the amounts typically used so he could vary the treatments around those levels.

CONTROL TREATMENTS

It is also necessary to include **control treatments** in an experiment. A control treatment is a treatment in which the independent variable is either eliminated or is set at a standard value. The results of the control treatment are compared to the results of the experimental treatments. In the fertilizer example, the investigator must be sure that the peanuts don't grow just as well with no fertilizer at all. The control would be a treatment in which no fertilizer is applied. An experiment on the effect of temperature on guinea pigs, however, cannot have a "no temperature" treatment. Instead, the scientist will use a standard temperature as the control and will compare weight gain at other temperatures to weight gain at the standard temperature.

For each of the following examples, tell what an appropriate control treatment would be.

1. An investigator studies the amount of alcohol produced by yeast when it is incubated with different types of sugars. Control treatment:

↳ The types of sugar / no sugar

2. The effect of light intensity on photosynthesis is measured by collecting oxygen produced by a plant. Control treatment:

↳ no light / natural light intensity

3. The effect of NutraSweet sweetener on tumor development in laboratory rats is investigated. Control treatment:

↳ no sweetener

4. Subjects are given squares of paper to taste that have been soaked in a bitter-tasting chemical. The investigator records whether each person can taste the chemical. Control treatment:

↳ don't soak in chemical

5. A solution is made up to simulate stomach acid at pH 2. Maalox antacid is added to the solution in small amounts, and the pH is measured after each addition. Control treatment:

↳ no antacid.

REPLICATION

Another essential aspect of experimental design is **replication**. Replicating the experiment means that the scientist repeats the experiment numerous times using exactly the same conditions to see if the results are consistent. Being able to replicate a result increases our confidence in it. However, we shouldn't expect to get exactly the same answer each time, because a certain amount of variation is normal in biological systems. Replicating the experiment lets us see how much variation there is and obtain an average result from different trials.

A concept related to replication is **sample size**. It is risky to draw conclusions based upon too few samples. For instance, suppose a scientist is testing the effects of fertilizer on peanut production. He plants four peanut plants and applies a different amount of fertilizer to each plant. Two of the plants die. Can he conclude that the amounts of fertilizer used on those plants were lethal? What other factors might have affected the results?

When you are designing experiments later in this lab course, consider sample size as an aspect of replication. Since there are no hard and fast rules to follow, seek advice from your instructor regarding the number of samples and the amount of replication that is appropriate for the type of experiment you are doing. Since the time you have to do experiments in lab is limited, inadequate replication may he a weakness of your investigations. Be sure to discuss this when you interpret your results.

METHODS

After formulating a hypothesis and selecting the independent and dependent variables, the investigator must find a method to measure the dependent variable; otherwise, there is no experiment. Methods are learned by reading articles published by other scientists and by talking to other scientists who are knowledgeable in the field. For example, a scientist who is testing the effect of fertilizer on peanuts would read about peanut growth and various factors that affect it. She would learn the accepted methods for evaluating peanut yield. She would also read about different types of fertilizers and their composition, their uses on different soil types, and methods of application. The scientist might also get in touch with other scientists who study peanuts and fertilizers and learn about their work. Scientists often do this by attending conferences where other scientists present results of investigations they have completed.

In this course, methods are described in the lab manual or may be learned from your instructor.

SUMMARY

Figure 1.1 summarizes the process of scientific investigation. The process begins and ends with the knowledge base, or what is already known. When a scientist chooses a new question to work on, he first searches the existing knowledge base to find out what information has already been published. Familiarity with the results of previous experiments as well as with the topic in general is essential for formulating a good hypothesis. After working through the rest of the process, the scientist contributes his own conclusions to the knowledge base by presentations at professional meetings and publication in scientific journals. Because each new experiment is built upon past results, the foundation of knowledge grows increasingly solid.

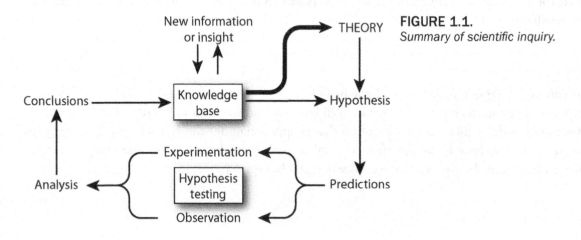

FIGURE 1.1.
Summary of scientific inquiry.

Scientific knowledge is thus an accumulation of evidence in support of hypotheses; it is not to be regarded as absolute truth. Hypotheses are accepted only on a trial basis. When you read about current scientific studies in the newspaper, keep in mind that the purpose of the media is to report news. In science, "news" is often preliminary results that are therefore quite tentative in nature. It is not unusual to hear that the results of one study contradict another. Some results will hold up under future scrutiny and some will not. However, this does not mean that scientific knowledge is flimsy and unreliable. All scientific knowledge falls somewhere along a continuum from tentative to certain, depending on the evidence that has been amassed. For example, it takes an average of 12 years to get a new drug approved by the FDA as researchers progress through laboratory evaluation of possible compounds, animal studies, and an escalating series of trials in humans. Even so, there are cases of drugs being recalled when new information is discovered. In a way, every medicine you take is still being tested—on you. We don't object to this because we feel confident that the knowledge base is firm, that the science is "done" to an acceptable degree.

Lab 1	
ACTIVITY 3	Designing an Experiment

9/15

OBJECTIVES

After completing this exercise, you should he able to

1. Given a proposed experiment, critique the experimental design.
2. Given a method for measuring a dependent variable, design an experiment.

Science is almost always a collaborative effort. Teams of scientists work together to solve problems; often these teams include scientists-in-training (graduate students), Working with others brings a variety of perspectives, knowledge bases, and experiences to bear on the problem. When scientists propose a project, they may seek funding from an agency such as the National Science Foundation. In this process, the team's proposal is reviewed by other scientists, who decide whether the problem is worth addressing, whether the proposers have the knowledge required to address it, and whether the design of the experiments is scientifically sound. When scientists have finished an investigation, they present the results to other scientists in the field by making oral presentations at conferences and by publishing articles in scientific journals.

Whether you are going to be a scientist or not, you will find many of the skills that scientists use are applicable to your own career. Most jobs require cooperative effort of some kind, just as you will collaborate with your lab team. Effective communication skills are especially important. You will almost certainly have occasion to present your work or defend your ideas to your coworkers and supervisor.

When you are asked to design an experiment in this course, your lab team will be provided with possible dependent variables and methods and procedures that you can use to measure them. You will decide what independent variable might affect these results. For example, if the topic being studied is the circulatory system, pulse rate and blood pressure might be the dependent variables that you measure to assess cardiovascular fitness. Your lab team would decide what factors (the independent variables) might affect a person's pulse rate and/or blood pressure and then design an experiment to test the effect of one of these factors.

Before you design your own experiments in later laboratories, you will work with your lab team in this part of the laboratory to critique a proposed experiment. This exercise is a rough draft of a proposal that could be improved. Use your knowledge of the scientific process to revise this experiment.

SAMPLE PROPOSAL

▶ Hypothesis

Athletes have better cardiovascular fitness than nonathletes.

▶ Dependent Variable(s)

Pulse rate, blood pressure.

Dickey, Jean, *Laboratory Investigations for Biology, 2nd ed.*, © 2003, pp. 1.4–1.14, A13–A24. Reprinted by permission of Pearson Education, Inc., Upper Saddle River, NJ.

▶ Independent Variable

Athletic training.

▶ Control(s)

Subjects who have had no athletic training (to have a comparison group of subjects); readings taken before exercise (to get a baseline measurement for each subject).

▶ Replication

Three subjects will be used in each group. Each subject will perform the exercise once.

▶ Brief Explanation of Experiment

The pulse rate and blood pressure of athletes and nonathletes will be measured. The subjects will then perform 5 minutes of aerobic exercise. The pulse rate and blood pressure of each subject will be measured again immediately after exercise.

▶ Predictions

We think that the pulse rates and blood pressure of athletes will be lower after exercise and will return to normal rates more quickly than those of nonathletes, indicating better cardiovascular fitness of athletes.

▶ Method

1. Recruit three athletes to be subjects. Our lab team will be the nonathlete subjects.
2. Record resting pulse rate and blood pressure for each subject.
3. All subjects will run up and down the stairs for 5 minutes.
4. Pulse rate and blood pressure of each subject will be measured immediately after the exercise.
5. Pulse rate will continue to be taken until it returns to the resting value. The time taken for each subject's pulse to return to normal will be recorded.
6. Measure and record blood pressure for each person when the resting pulse rate is reached.

Lab 2

GRAPHING AND SCIENTIFIC METHOD

PRE-LAB READING

Textbook: Appendix B: How to Interpret Graphs

Lab Manual: Lab 2.

ACTIVITIES

1. Data Presentation: Tables and Graphs
2. Graphing Practice
3. Interpreting Information on a Graph

REFERENCES

Dickey, J. 2003. *Laboratory Investigations for Biology.* 2nd edition. Benjamin Cummings, San Francisco, California.

Lab 2

ACTIVITY 1

Data Presentation: Tables and Graphs

This exercise explains how to

1. Distinguish between discrete and continuous variables.
2. Construct a line graph.
3. Construct a bar graph.
4. Choose the best method for presenting your data.

PART A: TABLES

A student team performed the experiment that was discussed in Lab 1. They tested the pulse and blood pressure of basketball players and nonathletes to compare cardiovascular fitness. They recorded the following data:

Nonathletes							Basketball Players					
	Resting Pulse			After Exercise				Resting Pulse			After Exercise		
	Trial			Trial				Trial			Trial		
Subject	1	2	3	1	2	3	Subject	1	2	3	1	2	3
1	72	68	71	145	152	139	1	67	71	70	136	133	134
2	65	63	72	142	144	158	2	73	71	70	141	144	142
3	63	68	70	140	147	144	3	72	74	73	152	146	149
4	70	72	72	133	134	145	4	75	70	72	156	151	151
5	75	76	77	149	152	153	5	78	72	76	156	150	155
6	75	75	71	154	148	147	6	74	75	75	149	146	146
7	71	68	73	142	145	150	7	68	69	69	132	140	136
8	68	70	66	135	137	135	8	70	71	70	151	148	146
9	78	75	80	160	155	153	9	73	77	76	138	152	147
10	73	75	74	142	146	140	10	72	68	64	153	155	155

If the data were presented to readers like this, they would see just lists of numbers and would have difficulty discovering any meaning in them. This is called **raw data**. It shows the data the team collected without any kind of summarization. Since the students had each subject perform the test three times, the data for each subject can be averaged. The other raw data sets obtained in the experiment would be treated in the same way.

TABLE. Average Pulse Rate for Each Subject
   (AVERAGE of 3 trials for each subject; pulse taken before and after 5-min step test)

	Nonathletes			Basketball Players	
	Resting Pulse	After Exercise		Resting Pulse	After Exercise
Subject	Average	Average	Subject	Average	Average
1	70	145	1	70	134
2	67	148	2	70	142
3	67	144	3	73	149
4	71	139	4	72	151
5	76	151	5	76	155
6	74	150	6	75	146
7	71	146	7	69	136
8	68	136	8	70	146
9	78	156	9	76	147
10	74	143	10	68	155

These rough data tables are still rather unwieldy and hard to interpret. A summary table could be used to convey the overall averages for each part of the experiment. For example:

TABLE. Overall Averages of Pulse Rate
   (10 subjects in each group; 3 trials for each subject;
   pulse taken before and after 5-min step test)

	Pulse Rate (beats/min)	
	Before Exercise	After Exercise
Nonathletes	71.6	145.8
Basketball players	71.9	146.1

Notice that the table has a title above it that describes its contents, including the experimental conditions and the number of subjects and replications that were used to calculate the averages. In the table itself, the units of the dependent variable (pulse rate) are given and the independent variable (nonathletes and basketball players) is written on the left side of the table.

Tables should be used to present results that have relatively few data points. Tables are also useful to display several dependent variables at the same time. For example, average pulse rate before and after exercise, average blood pressure before and after exercise, and recovery time could all be put in one table.

PART B: GRAPHS

Numerical results of an experiment are often presented in a graph rather than a table. A graph is literally a picture of the results, so a graph can often be more easily interpreted than a table. Generally, the independent variable is graphed on the x axis (horizontal axis) and the dependent variable is graphed on the y axis (vertical axis). In looking at a graph, then, the effect that the independent variable has on the dependent variable can be determined. "Dry mix" may help you remember which variable goes on which axis. The **d**ependent variable is sometimes called the **r**esponding variable and it goes on the *y* axis. **M**anipulated variable is a term sometimes used for the **i**ndependent variable, and it goes on the *x* axis.

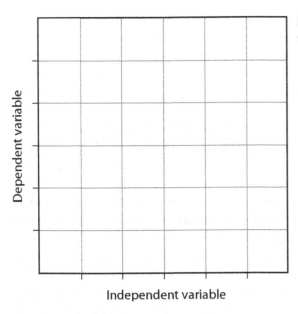

FIGURE 2.1.
Graph construction.

When you are drawing a graph, keep in mind that your objective is to show the data in the clearest, most readable form possible. In order to achieve this, you should observe the following rules:

- Use graph paper to plot the values accurately.
- Plot the independent variable on the *x* axis and the dependent variable on the *y* axis. For example, if you are graphing the effect of the amount of fertilizer on peanut weight, the amount of fertilizer is plotted on the *x* axis and peanut weight is plotted on the *y* axis.
- Label each axis with the name of the variable and specify the units used to measure it. For example, the *x* axis might be labeled "Fertilizer applied (g/100 m²)" and the *y* axis might be labeled "Weight of peanuts per plant (grams)."

- The intervals labeled on each axis should he appropriate for the range of data so that most of the area of the graph can be used. For example, if the highest data point is 47, the highest value labeled on the axis might be 50. If you labeled intervals on up to 100, there would be a large unused area of the graph.

- The intervals that are labeled on the graph should be evenly spaced. For example, if the values range from 0 to 50, you might label the axis at 0, 10, 20, 30, 40, and 50. It would be confusing to have labels that correspond to the actual data points (for example, 2, 17, 24, 30, 42, and 47).

- The graph should have a title that, like the title of a table, describes the experimental conditions that produced the data.

Figure 2.2 illustrates a well-executed graph.

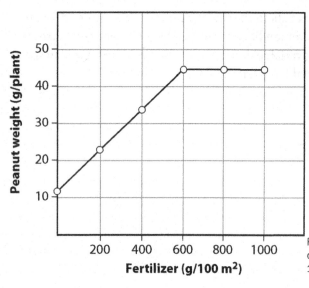

FIGURE 2.2.
*Graph of peanut weight vs. amount
of fertilizer applied.*

Figure 1. Weight of peanuts produced per plant when amount
of fertilizer applied is varied. (Average seed weight per plant in
100 m² plots, 400 plants/plot.)

The most commonly used forms of graphs are line graphs and bar graphs. The choice of graph type depends on the nature of the independent variable being graphed. **Continuous variables** are those that have an unlimited number of values between points. **Line graphs** are used to represent continuous data. For instance, time is a continuous variable over which things such as growth will vary. Although the units on the axis can be minutes, hours, days, months, or even years, values can be placed in between any two values. Amount of fertilizer can also be a continuous variable. Although the intervals labeled on the x axis are 0, 200, 400, 600, 800, and 1,000 g/100 m², many other values can be listed between each two intervals.

In a line graph, data are plotted as separate point the axes, and the points are connected to each other. Notice in Figure 2.3 that when there is more than one set of data on a graph, it is necessary to provide a key indicating which line corresponds to which data set.

Discrete variables, on the other hand, have a limited number of possible values, and no values can fall between them. For example, the type of fertilizer is a discrete variable: There are a certain number of types which are distinct from each other. If fertilizer type is the independent variable displayed on the x axis, there is no continuity between the values.

Bar graphs, as shown in Figure 2.4, are used to display discrete data.

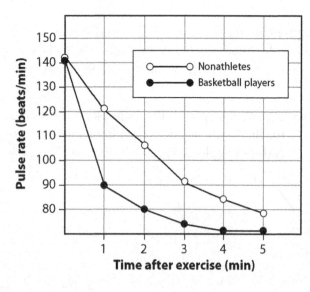

FIGURE 2.3.
*Line graph representing two
related sets of data.*

Figure 1. Recovery rate of basketball players and nonathletes
after performing a step test for 5 minutes. (Average of 10
subjects; each subject performed the test 3 times.)

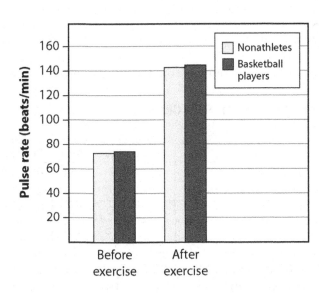

FIGURE 2.4.
Example of bar graph.

Figure 1. Average pulse rates of basketball players and nonathletes before and after performing a step test for 5 minutes. (Average of 10 subjects; each subject performed the test 3 times.)

In this example, before- and after-exercise data are discrete: There is no possibility of intermediate values. The subjects used (basketball players and nonathletes) also are a discrete variable (a person belongs to one group or the other).

This graph could also have been constructed as shown in Figure 2.5.

What is the difference between the two graphs?

Explain why the first way would be better to convey the results of the experiment.

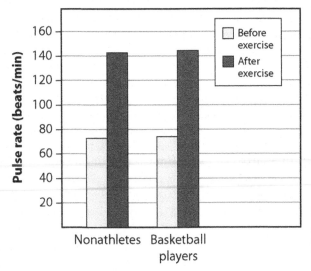

FIGURE 2.5.
Alternative method of presenting data in Figure 2.4.

Figure 1. Average pulse rates of basketball players and nonathletes before and after performing a step test for 5 minutes. (Average of 10 subjects; each subject performed the test 3 times.)

ACTIVITY 2: GRAPHING PRACTICE

Use the temperature and precipitation data provided in Table 2.1 to complete the following questions:

TABLE 2.1. Average Monthly High Temperature and Precipitation for Four Cities

(T = temperature in °C; P = precipitation in cm)

		Jan.	Feb.	Mar.	Apr.	May	June	July	Aug.	Sept.	Oct.	Nov.	Dec.
Fairbanks, Alaska	T	−19	−12	−5	6	15	22	22	19	12	2	−11	−17
	P	2.3	1.3	1.8	0.8	1.5	3.3	4.8	5.3	3.3	2.0	1.8	1.5
San Francisco, California	T	13	15	16	17	17	19	18	18	21	20	17	14
	P	11.9	9.7	7.9	3.8	1.8	0.3	0	0	0.8	2.5	6.4	11.2
San Salvador, El Salvador	T	32	33	34	34	33	31	32	32	31	31	31	32
	P	0.8	0.5	1.0	4.3	19.6	32.8	29.2	29.7	30.7	24.1	4.1	1.0
Indianapolis, Indiana	T	2	4	9	16	22	28	30	29	25	18	10	4
	P	7.6	6.9	10.2	9.1	9.9	10.2	9.9	8.4	8.1	7.1	8.4	7.6

Source: Pearce, E. A., and G. Smith. Adapted from *The Times Books World Weather Guide.* New York: Times Books, 1990.

1. Compare monthly temperatures in Fairbanks with temperatures in San Salvador.

   Can data for both cities be plotted on the same graph?

   What will go on the *x* axis?

   How should the *x* axis be labeled?

   What should go on the *y* axis?

   What is the range of values on the *y* axis?

   How should the *y* axis be labeled?

   What type of graph should be used?

2. Compare the average September temperature for Fairbanks, San Francisco, San Salvador, and Indianapolis.

   Can data for all four cities be plotted on the same graph?

Lab 3

EARTH STRUCTURE AND PLATE TECTONICS

PRE-LAB READING

Textbook: Chapter 2: Geology.

Lab Manual: Lab 3.

ACTIVITIES

1. Modeling Convection
2. Evidence of Pangaea
3. OmniGlobe

INTRODUCTION

The origin of the physical features of the earth's surface that we see and experience throughout our lives as land-dwelling humans can be traced back to the movements of the earth's tectonic plates. And the movement of the tectonic plates is driven by and dependent on structures and processes related to the interior structure of our planet.

The most basic description of the earth's structure identifies three layers: the crust, the mantle, and the core. The outermost layer of the earth is the **crust**. It is a very thin layer of rigid material that interfaces with the atmosphere. Just below the crust is a thick region called the **mantle**. The physical properties of the mantle vary. The uppermost portion of the mantle, just below the crust, is rigid. But below that, there is a portion of the mantle that is soft and bendable (plastic). And below that, the mantle is solid. Finally, at the center of the earth is the **core**, part of which is liquid and part of which is solid. The core is responsible for generating heat and the earth's magnetic field.

Because of the variations within the three layers identified in the most basic description of the earth's structure, it is useful to use a more detailed description of the earth's structure.

There are two types of crust; oceanic and continental. **Oceanic crust** is crust that lies beneath the oceans and **continental crust** forms the continents. Oceanic crust is thinner and denser than continental crust.

The uppermost portion of the **mantle**, just under the crust, is rigid and this and the crust above it together form the **lithosphere**. **Tectonic plates** consist of pieces of the lithosphere. The next portion of the mantle, the soft, bendable (plastic) portion is called the **asthenosphere**. Heat from the core causes convection currents in this part of the mantle. The convection is responsible for the movement of tectonic plates. The remainder of the mantle, between the asthenosphere and outer core, is solid and is called the **mesosphere**.

The earth's **core** consists mostly of nickel and iron and it can be divided into the outer core and the inner core. The **outer core** is the outermost portion of the core and is adjacent to the mesosphere of the mantle. The outer core is liquid (molten). Currents in the liquid outer core are responsible for the earth's magnetic field. The **inner core** is solid.

The temperature, pressure, and density of the earth's interior layers increase with increases in depth.

The lithosphere of the earth consists of about 15 major **tectonic plates**. These plates are in constant motion as a result of the convection currents in the asthenosphere of the mantle. Interactions between these moving plates are responsible for many of the physical features and processes that can be observed on the surface of the earth, such as earthquakes and volcanoes. Adjacent plates are separated from one another at **plate boundaries**.

There are three main types of plate boundaries: Convergent, divergent, and transform.

Convergent boundaries are places where two adjacent plates are moving *toward* each other. If one of the plates is denser than the other, when the two plates collide, the denser of the two will be dragged down under the less dense in a process called **subduction**. This scenario is typical during the convergence of a dense oceanic plate (oceanic crust plus uppermost mantle) and a less dense continental plate (continental crust plus uppermost mantle). Earthquakes and volcanoes can occur in areas undergoing subduction. If the two converging plates are of equal density then neither will be subducted under the other. Instead, a mountain chain will be formed.

Divergent boundaries are places where two adjacent plates are moving *away from* each other. As two plates diverge, **magma** (molten material from the mantle) rises to the surface where it cools and solidifies upon contact with water (as in oceans) or air and forms new crust. At divergent boundaries under oceans this process is called **seafloor spreading**.

Transform boundaries are places where two adjacent plates are sliding past one another. As the plates slide past one another they become caught up or locked against one another, causing tension to build up. When that tension is eventually released, when the plates become unstuck or unlocked, earthquakes occur.

In this exercise, you will carry out several activities to model and illustrate the interior structure of the earth and plate tectonics.

REFERENCES

Anonymous. 2006. Caroline modeling tectonic plate boundaries kit. Carolina Biological Supply Company, Burlington, North Carolina.

Anonymous. 2006. Evidence of Pangaea Kit. Carolina Biological Supply Company, Burlington, North Carolina.

Padilla, M. J., M. Cyr, J. D. Exline, I. Miaoulis, J. M. Pasachoff, B. B. Simons, C. G. Vogel, and T. R. Wellnitz. 2001. Science explorer: earth science. Prentice- Hall, Inc., Upper Saddle River, New Jersey.

Robertson, Eugene C. 2007. The interior of the earth. *United States Geologic Survey.* Retrieved May 7, 2010. From http://pubs.usgs.gov/gip/interior/.

Steer, D.N., C.C. Knight, K.D. Owens, and D.A. McConnell. 2005. Challenging students ideas about earth's interior structure using a model-based, conceptual change approach in a large class setting. Journal of Geoscience Education, 53(4):415-421. Retrieved May 7, 2010. From http://repository.lib.ncsu.edu/publications/bitstream/1840.2/1601/1/mcconnell+2.pdf.

Withgott, Jay and S. Brennan. 2008. Environment: the science behind the stories, 3rd edition. Pearson Education Inc., California.

Lab 3
ACTIVITY 1 Modeling Convection

OBJECTIVES

- Understand the role of heat in developing convective circulation.
- Understand convective circulation and convective cells.
- Understand that the source of the heat driving convection within the earth is the earth's core.
- Observe convective circulation and convective cells in a closed system.

HYPOTHESES

- Convective circulation will develop in the convection fluid when the bottle containing it is subjected to a heat source.

MATERIALS

- Bottle, French square, glass, ~225 mL capacity, 1
- Blocks, wooden blocks, (size sufficient to raise bottle/pan above candle), ≥4
- Candle, small (votive size), 3
- Candle saucer, 1
- Convection fluid, ~225 mL
- Flashlight, 1
- Matches (or candle lighter), 1
- Pan, aluminum or stainless steel, ~15 cm × 30 cm and ~4 cm deep, 1

PROCEDURE

▶ Modeling Convection Within the Earth

1. You have been provided a square bottle containing convection fluid. The bottle's lid has been sealed with tape. Do not remove the tape or open the bottle.
2. Shake the bottle of convection fluid vigorously to suspend all of the material.

3. Elevate both ends of the bottle with wooden blocks, as illustrated in Figure 3.1.

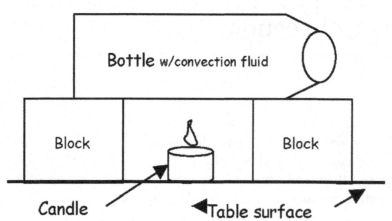

FIGURE 3.1.
Illustration of the placement and elevation of convection fluid bottle on wooden blocks and the placement of candle under the bottle.

4. Place the candle in the saucer-like container (whatever has been provided by your instructor), which will catch any melted wax and keep it from spilling onto the lab table.

5. Light the candle, using matches or a candle lighter, and slide it into place between the blocks and under the bottle (as illustrated in Figure 3.1). **CAUTION**: Be extremely careful around the open flame of the candle! Keep paper, hair, clothes, wooden blocks and any other flammable material out of direct contact with the flame. If a match was used to light the candle, make sure it is extinguished and cool before discarding it.

6. Allow the candle to burn under the bottle for about 3–5 minutes before making your observations.

7. Position yourself so that your eyes are level with the side of the bottle. From this position, observe the movement of the convection fluid within the bottle. You should be able to detect in the fluid the pattern of movement characteristic of convective circulation. If you have difficulty detecting the movement, shine the flashlight down on the bottle from above while viewing it from the side.

8. Record on Figure 3.2 on the data sheet, arrows that illustrate the pattern of movement you observed in the fluid while viewing the bottle from the *side*.

9. Now, observe the movement of the convection fluid within the bottle when viewing it while looking down on it from above. If you have difficulty detecting the movement, shine the flashlight on the bottle from the side while viewing it from the top.

10. Record in Figure 3.3 on the data sheet, arrows that illustrate the pattern of movement you observed in the fluid while viewing the bottle from *above*.

11. **CLEANUP:** When you are finished, make sure the following cleanup steps have been carried out:
    a. **CAUTION:** The bottle of convection fluid may still be hot when you cleanup so be careful!
    b. Wipe any soot from the bottle with a paper towel.
    c. Make sure all candles have been extinguished and have cooled before placing them in any storage container.
    d. If any wax has spilled on the lab table, clean it up.

Lab 3

ACTIVITY 1

Modeling Convection

Student Name: _____   Lab Group:_____

TA: _____   Lab Date/Section: _____

DATA

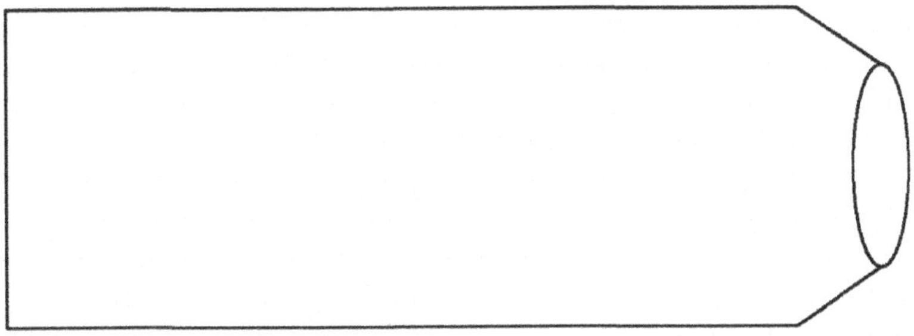

FIGURE 3.2.
*Diagram of movement (arrows) of convection fluid in bottle placed over a heat source, as viewed from the **side**.*

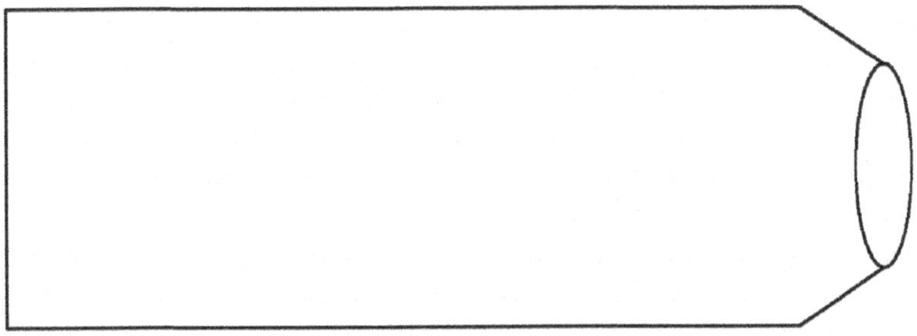

FIGURE 3.3.
*Diagram of movement (arrows) of convection fluid in bottle placed over a heat source, as viewed from the **top**.*

Lab 3

ACTIVITY 1

Modeling Convection

Student Name: _____   Lab Group:_____

TA: _____   Lab Date/Section: _____

DISCUSSION & CONCLUSIONS

For full credit, questions must be answered thoroughly, in complete sentences.

1. *Describe* the motion of the convection fluid as viewed from the side.

   _____

   _____

   _____

2. *What* was driving the motion of the convection fluid in the *bottle*?

   _____

   _____

   _____

3. *What* is the ultimate source of the heat that drives convective circulation in the mantle?

   _____

   _____

4. *Explain* convective circulation in a convective cell.

   _____

   _____

   _____

   _____

5. *What* are tectonic plates composed of?

   _____

   _____

   _____

For the hypothesis below, state whether or not it was supported by the data collected and explain your response.

> *Hypothesis:* Convective circulation will develop in the convection fluid when the bottle containing it is subjected to a heat source.

_____

_____

_____

_____

_____

_____

_____

_____

_____

_____

_____

_____

_____

_____

_____

Lab 3
ACTIVITY 2 — Evidence of Pangaea

OBJECTIVES

▪ Learn the categories of evidence used to help develop the concepts of Pangaea and the theory of plate tectonics.
▪ Understand how specific examples of evidence line up across the boundaries of present-day continents when they are placed in their Pangaea-era configuration.

HYPOTHESIS

▪ None.

MATERIALS

▪ Magnet-receptive map board (from "Pangaea Puzzle Kit" by Carolina Biological Supply)
▪ Magnetic cutouts of earth's continents/land masses (from "Pangaea Puzzle Kit" by Carolina Biological Supply) labeled with symbols representing evidence for Pangaea

PROCEDURE

1. Obtain a magnet-receptive map board and set of continent/land mass magnets. There may be one set of these for each lab table or there may only be a few which must be shared among all the lab groups. If the set(s) must be shared, retrieve a set from the common area and return it there when you are finished with it.
2. The set of continent/land mass magnets represent nine different continents/land masses: North America, South America, Africa, Madagascar, Greenland, Eurasia, India, Antarctica, and Australia. Place the magnets in their present day positions on the map board. The shapes of the magnets are not exactly the same shape as the outlines of the present day continents on the map board, so take care as you decide where to place the magnets and ask your instructor for assistance if you are confused or uncertain.

3. You will now "construct" Pangaea using the continent/land mass magnets. **IMPORTANT:** Start by placing the Antarctica magnet ~2 cm north of the center of the bottom edge of the map board and fit together all the other continents/land masses using Antarctica (in the position described) as your anchor point.

4. Refer to the map, magnets, and background information in Table 3.1 on the discussion and conclusions sheet to answer the questions on the same sheet.

5. **CLEANUP:** When you are finished, make sure the following cleanup steps have been carried out:
   a. Deconstruct Pangaea and arrange the continent/land mass magnets randomly on the map board.
   b. If the map board and continent/land mass magnet sets are being shared, return yours to the common area.

Lab 3

ACTIVITY 2 Evidence of Pangaea

Student Name: _____   Lab Group:_____

TA: _____   Lab Date/Section: _____

DISCUSSION & CONCLUSIONS

For full credit, questions must be answered thoroughly, in complete sentences.

TABLE 3.1. Background Information about the Categories of Evidence, Along with the Legend on the Map Board, Needed to Answer the Questions Below

Fossils	Physical/Geologic Features
Glossopteris: a fern-like plant that lived ~260 million years ago.	**Glacial scarring:** occurs in areas that are or were covered by glaciers.
Cynognathus: a reptilian animal that lived ~240 million years ago.	**Mountain ranges:** those represented on the magnetic continents/land masses are of the same age and geologic structure.
Lystrosaurus: a reptilian animal that lived ~240 million years ago.	**Subtropical deposits:** coal deposits formed in areas where the remains of subtropical (and tropical) plants accumulated.
Mesosaurus: a reptilian animal that lived ~260 million years ago.	

1. *Which* portions (i.e., northern, east edge) of which continents/land masses contain fossils of *Glossopteris*? *Do* these areas of the continents/land masses line up when the continents/land masses are arranged as Pangaea?

   _____

   _____

   _____

   _____

2. Which portions (i.e., northern, east edge) of which continents/land masses contain fossils of *Cynognathus*? Do these areas of the continents/land masses line up when the continents/land masses are arranged as Pangaea?

   _____

   _____

   _____

3. Which portions (i.e., northern, east edge) of which continents/land masses contain fossils of *Lystrosaurus*? Do these areas of the continents/land masses line up when the continents/land masses are arranged as Pangaea?

_____

_____

_____

4. Which portions (i.e., northern, east edge) of which continents/land masses contain fossils of *Mesosaurus*? Do these areas of the continents/land masses line up when the continents/land masses are arranged as Pangaea?

_____

_____

_____

5. Which, if any, of the locations of the four types of fossils are inconsistent with the general climate that would have been expected in those areas of Pangaea?

_____

_____

_____

6. Which portions (i.e., northern, east edge) of which continents/land masses contain glacial scarring? Do these areas of the continents/land masses line up when the continents/land masses are arranged as Pangaea?

_____

_____

_____

7. Is the pattern of glacial scarring on Pangaea consistent with the general climate that would have been expected in those areas of Pangaea? Why?

_____

_____

_____

8. Which portions (i.e., northern, east edge) of which continents/land masses contain mountain ranges? Do these areas of the continents/land masses line up when the continents/land masses are arranged as Pangaea?

_____

_____

_____

9. Which portions (i.e., northern, east edge) of which continents/land masses contain subtropical deposits? Do the areas of the continents/land masses line up when the continents/land masses are arranged as Pangaea?

   _____

   _____

   _____

10. Which continent/land mass has moved the farthest from its position in Pangaea to its present day position?

    _____

11. Which continent/land mass has moved the least from its position in Pangaea to its present day position?

    _____

12. What are the categories of evidence that helped support the concept of the existence of the super-continent Pangaea?

    _____

    _____

Lab 3
ACTIVITY 3 OmniGlobe

PROCEDURE

1. Follow instructions of your TA, collect your belongings in the lab and go to the lobby of Oceanography building to meet your group at the OmniGlobe.
2. Observe the animation of plates tectonics through geological time, map of the age of sea floor, cumulative earthquake activities, and map of world vulcanoes.
3. Answer the questions on the discussion and conclusions sheet.

Lab 3
ACTIVITY 3
OmniGlobe

Student Name: _____   Lab Group:_____

TA: _____   Lab Date/Section: _____

DISCUSSION & CONCLUSIONS

For full credit, questions must be answered thoroughly, in complete sentences.

1. At what time did the large southern landmass called Pannotia begin to break apart into several small pieces?

   _____

2. Which mountain range is the highest on Earth and why?

   _____

   _____

   _____

3. Where is "The Ring of Fire" located and why is it named like that?

   _____

   _____

   _____

4. Where do most earthquakes and volcanoes occur in relation to tectonic plate boundaries?

   _____

   _____

   _____

Lab 4

DARWINIAN SNAILS

PPE-LAB READING

Textbook: Chapter 3: Evolution.

Lab Manual: Lab 4.

OBJECTIVES

- Understand the assumptions behind the Darwin's theory of evolution by natural selection—variation among individuals, heritability, and differential reproduction—as well as the role of mutation in creating genetic variation.
- Use a simulation model to "violate" each assumption and explore the conditions under which evolution occurs.
- Understand how model systems can be used in scientific research of evolution.
- Practice to plot and interpret graphs.

MATERIAL

- SimBio Virtual Labs® EvoBeaker®: Darwinian Snails.

ACTIVITIES

1. A Model of Evolution by Natural Selection
2. The Requirements for Evolution by Natural Selection
3. Darwin's Theory of Evolution by Natural Selection

INTRODUCTION

The flat periwinkle is a small snail that lives on seaweeds growing on rocky shores in New England. Among the snail's enemies is the European green crab. As its name suggests, the European green crab is not native to North America. It traveled from Europe early in the 19th century. Before 1900, the green crab did not occur north of Cape Cod, Massachusetts. After the turn of the century, however, the crab expanded its range northward, and is now found as far north as Nova Scotia. The crab's range expansion introduced periwinkle populations north of Cape Cod to a new predator.

Biologist Robin Seeley suspected that New England's periwinkle populations have evolved due to predation by green crabs. In a museum, Seeley found an 1871 collection of periwinkles from Appledore Island, north of Cape Cod. She compared these old shells to new shells she had gathered herself at the same place. Seeley measured the thickness of each shell. Her data appear in the table below.

SHELLS COLLECTED IN 1982–1984, EIGHTY YEARS AFTER EUROPEAN GREEN CRABS ARRIVED

14	11	12	14	9	11	12	13	14	10	14	10	14	12	13
10	15	13	14	15	15	13	15	11	16	12	18	16	16	17
17	17	15	12	13	15									

SHELLS COLLECTED IN 1871, BEFORE EUROPEAN GREEN CRABS ARRIVED

5	8	1	9	4	8	8	8	11	3	9	8	7	8	5
8	8	13	9	10	9	5	2	8	7	5	9	8	8	5
11	12	8	3	3	5	9	8	6	7					

Each number represents a single shell. (Don't worry about the unit of measure. Seeley plotted her data as the logarithm of thickness, which we have converted to integers.) Bigger numbers mean thicker shells.

Plot the data for the shells collected in the 1980s on the top grid at the right. If you're working with a partner, one of you should read the number while the other places an X on the grid for each shell. We have put an X on the grid for the first shell (with a thickness of 14); you add the rest. When you have more than one shell with the same thickness, stack the X's on top of each other.

Next, plot the data for the shells collected in 1871 on the bottom grid at the right.

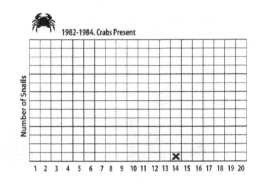

The graphs you have just drawn are called histograms. Histograms are a powerful way to summarize the variation among individuals in a population, and to compare populations to each other.

On the top histogram you just drew, showing the periwinkles from the 1980s, **mark the average shell thickness of the snail population with a triangle as we have done at the right.** Don't do any calculations; just judge the average by eye— it is the point at which your collection of X's would balance. **Also mark the average thickness of the 1871 periwinkle population.**

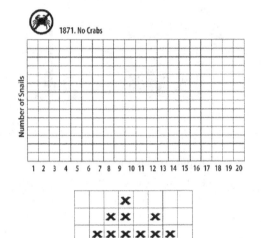

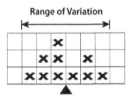

Average Shell Thickness

On the histogram showing the periwinkles from the 1980s, mark the range of variation in thickness among the snails with a two-headed arrow as we have done at right. For the collection of X's at right, the range of variation is 6 units. Also mark the range of variation in thickness among the periwinkles from 1871.

Note that the average shell thickness and the range of variation in shell thickness are properties of a population of snails, not properties of any single individual. There may not be any individual snails whose shell thickness is exactly equal to the average, and no individual snail has a range of variation in thickness. Only a group of snails, taken together, can have an average and a range of variation.

Recall that Robin Seeley had predicted that the flat periwinkle population on Appledore Island would have evolved between 1871 and the 1980s. Like average and range of variation, evolution is a term that applies to a population of organisms, not to any single individual. Evolution marks a change in the composition of the population; it is typically reflected in a change in the average and/or the range of variation.

Compare your histograms to the histograms in the illustration below (reprinted from Seeley, 1986), which we have drawn from Seeley's data. As the graphs and photos in the illustration show, the snail population on Appledore Island in the early 1980s was, indeed, dramatically different from the snail population that was there in 1871. The snails had, on average, shells that were thicker than those of their ancestors. The 1980s population also showed a somewhat smaller range of variation in shell thickness. The flat periwinkles living on Appledore in the early 1980s were descendants of the snails that were living there in 1871. Therefore, we can describe the change in the population as descent with modification, or evolution.

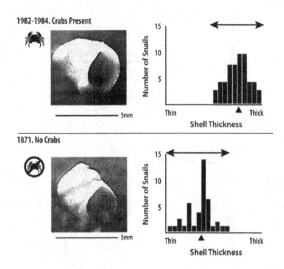

How did this descent with modification, this evolution, happen? The mechanism of evolution is the subject of this lab. You will do experiments on a model population to explore how evolution works. Then you will return to Seeley's flat periwinkles to see how the model applies to them.

REFERENCES

Seeley, R. H. 1986. Intense natural selection caused a rapid morphological transition in a living marine snail. *Proceedings of the National Academy of Sciences, USA* 83: 6897-6901.

Trussell, G. C. 1996. Phenotypic plasticity in an intertidal snail: The role of a common crab predator. *Evolution 50:* 448-45.

Lab 4
ACTIVITY 1
A Model of Evolution by Natural Selection

PROCEDURE

1. Launch the **SimBio Virtual Labs** program. Select **Darwinian Snails** from the EvoBeaker Labs option.

2. You will see a population of snails scattered around the Rocky Coastline on the left. Take a closer look at these snails by double-clicking on one of the snails. A window will pop up showing you an enlarged view of that snail and the thickness of its shell. Examine at least 9 other snails in this way.

   2.1 Record the shell thickness on the Data sheet and answer the question 1.

3. Look at the histogram on the right side of the screen. This sorts the snails on the coastline by shell thickness, and shows the number of snails in each category.

   3.1 Answer question 2 on the Data sheet.

   You will now become a European green crab. You will feel especially crabby if you are not getting enough to eat, and the best snacks available on the coastline are these tasty looking snails. All you have to do is crack their shells by pounding on them with your claw.

4. Before beginning your feast, copy the histogram of shell thicknesses and save it in a text document. To do this, move your mouse to the center of the histogram, right-click (Windows) or Control-click (OSX) and select "Copy View to Clipboard". Finally, open a new document in your word processor and use the paste command to paste the histogram into the document. Label this graph **Activity 1: Starting Population** so that you remember what the snails looked like when you first got to the Coastline.

   4.1 Answer question 1 on Discussion & Conclusions sheet.

5. Begin the simulation by clicking the **GO** button (the left-most button in the Controls panel). The snails will start to crawl around. The starting population size is 50, as shown by the Current Snail Population item below the Coastline.

6. Get your claw ready for action by clicking on the **CLAW** tool (the crab claw button in the Tools panel). You are now officially a European green crab.

7. Find a snail you want to eat and start clicking on it. When you claw at it enough times, the shell will crack, you'll eat what's inside, and the snail will disappear. The **Current Snail Population** will also show that there is one fewer snail on the Coastline.

8. Notice the **Crab Happiness Score** below the Coastline. This score will go up every time you eat a snail, but it will go down every time you click on a snail with your claw, because the more effort you expend to get your meal, the 'crabbier' you become. The coastline started with 50 snails, and it takes 25 snails to fill you up. Keep eating snails until you have eaten 25 (with 25 remaining), and try to maximize your **Crab Happiness Score** while doing this.

   8.1. Answer question 2 on Discussion & Conclusions sheet.

9. You, the crab, go away for a while after your big meal, and the snails have a chance to reproduce. Stop the model by clicking the **STOP** button (the square button in the Controls panel). Then allow the snails to reproduce by clicking the **REPRODUCE** button.

   Each of the surviving snails generates two new snails by cloning, then dies. Unlike with most real snails, there is no mating. Each offspring is identical to its parent. To see this, switch back to the **SELECT** tool (the standard arrow button in the Tools panel) and double-click on a few pairs of nearby snails (twin children of a single parent). You should see that the shell thickness is the same in both snails in each pair.

10. You are hungry again! Start the simulation by clicking the **GO** button, click on the **CLAW** tool so you can eat again, and have another meal of 25 snails. Don't forget to try to keep your **Crab Happiness Score** high by eating your snails with the fewest clicks possible.

11. Digest your meal and let the snails reproduce again by clicking on the **REPRODUCE** button. Then have one more meal of 25 snails. Yum! At the end, **REPRODUCE** once more so you finish with about 50 snails.

12. When your third meal is done, **copy** the histogram of shell thicknesses and paste it into your text document. Label it **Activity 1: Generation 4**. (Remember, to copy a histogram, mouse over the center of the histogram, right-click (Windows) or Control-click (OSX), and select "Copy View to Clipboard". Then use the paste command to place the screen shot in your word processing document.)

   12.1.   Answer question 3 on Discussion & Conclusions sheet.

13. Playing crab may become tedious after a while. To help with this, you can add auto-munching crabs to the Rocky Coastline. These crabs will eat snails automatically while you watch. To do this, first click **RESET** to start over with a new population of snails. **Copy** the histogram of shell thicknesses and paste it into your text document. Label it **Activity 1: Before Crabs**.

14. Now click on the **ADD CRITTER** tool in the Tools panel (just to the right of the **CLAW** button in the Tools panel). Hold the mouse down on the button's pop-up arrow and select the crab icon. Then, click among the snails on the Coastline. Each click will add a crab. Add three to five crabs to the Coastline.

15. Run the simulation (with the **GO** button) and watch the crabs as they eat snails. Keep an eye on the size of the snail population. When there are about 25 snails left, **STOP** the model.

16. **Copy** the histogram of shell thickness and paste it into your text document. Label it **Activity 1: After Crabs**. Compare your before-crabs and after-crabs histograms.

   16.1.   Answer question 4 on Discussion & Conclusions sheet.

17. Click the **REPRODUCE** button to let the snails reproduce. Run the simulation again until only 25 snails are left, then stop it. Click the **REPRODUCE** button again, run the simulation until only 25 snails are left, and then stop it. Click the **REPRODUCE** button once more.

   17.1.   Answer question 5 on Discussion & Conclusions sheet.

Lab 4

ACTIVITY 2

The Requirements for Evolution by Natural Selection

In Activity 1 you saw that snail population evolve (change over time) as a result of predation—a form of selection. What is required for such change to take place? This activity explores several conditions that affect whether the population will change as individuals are selected.

▶ Part 1. Variation

PROCEDURE

1. What if all the snails started out the same? Could the population still evolve? To get a uniform population of snails with no variation in shell thickness, click the '**Shell thickness is variable**' checkbox so that it is unchecked.
2. To get a new population of snails, reset the model by clicking on the **RESET** button.
3. Click the **SELECT** (arrow) tool. Look at the shell thickness of a few snails. Also look at the histogram of shell thickness.

   3.1. Answer question 1 on Data sheet.

4. **Copy** the histogram and paste it into your text document, and label it **Activity 2: No Variation, Starting Population**.

   4.1. Answer question 1 on Discussion & Conclusions sheet.

5. Now test your hypothesis. Click on the **ADD CRITTER** tool and select the crab icon. Add three to five crabs to the Rocky Coastline. Run the simulation (with the GO button) and let the crabs eat 25 snails.
6. **STOP** the model and let the snails **REPRODUCE**. **COPY** the histogram from the screen, paste it into your text document, and label it **Activity 2: No Variation, Generation 2**.

   6.1. Answer question 2 on Discussion & Conclusions sheet.

▶ Part 2. Inheritance

PROCEDURE

7. In the Activity 1 experiments, the shell thickness of each snail was identical to that of its parent. Shell thickness was completely genetically determined, so each snail inherited the genes for shell thickness from its parent and so its shell was exactly the same thickness as the parent's. What if this weren't so? What if shell thickness was not heritable—if there was no genetic basis for shell thickness—but each snail instead grew its shell to a random thickness that had nothing to do with its parent's shell? To find out, click the '**Shell thickness is heritable**' box so it is unchecked.

8. Reestablish variation in shell thickness in the model by clicking on the '**Shell thickness is variable**' checkbox. This should now be checked so that variation is present.

9. **RESET** the model. Before starting the simulation, see what will happen during reproduction by clicking on the **REPRODUCE** button. Click the **SELECT** tool and look at a few pairs of snails.

   9.1. Answer question 3 on Discussion & Conclusions sheet.

   9.2 Answer question 4 on Discussion & Conclusions sheet.

10. **RESET** the model again so that you start with 50 snails as before. **COPY** the histogram from the screen and label it **Activity 2: No Inheritance, Starting Population**.

11. Click on the **ADD CRITTER** tool and add three to five crabs. Run the simulation until there are just 25 snails left.

12. **REPRODUCE**, and then let your crabs have two more meals, making sure to **REPRODUCE** after each one.

13. **COPY** the new shell thickness distribution from the screen, paste it into your text document, and label it **Activity 2: No Inheritance, Generation 4**.

   13.1.   Answer question 5 on Discussion & Conclusions sheet.

▶ Part 3. Selection

PROCEDURE

14. Imagine a crab that is especially large and can crack snail shells no matter how thick they are. This crab just takes snails to eat randomly, without any preference for thinner shells. Will there still be a change in the distribution of shell thickness over time? To test this, click on the '**Survival is selective**' checkbox so it is unchecked.

15. Make shell thickness heritable again by clicking on the '**Shell thickness is heritable**' checkbox so that it is checked. To run this experiment, '**Survival is selective**' should be unchecked while '**Shell thickness is variable**' and 'Shell thickness is heritable' should be checked. You should also see a button called **EAT RANDOM SNAIL** that has become active next to the **REPRODUCE** button.

16. **RESET** the model and click **GO**. Do not add crabs. Click the **EAT RANDOM SNAIL** button a few times and watch what is happening to the snails on the Coastline. You will see that each time you click **EAT RANDOM SNAIL**, one randomly chosen snail disappears. You'll see both thin-shelled and thick-shelled snails disappearing with equal probability.

17. **RESET** the model. **COPY** the histogram from the screen, paste it into your text document, and label it **Activity 2: No Selection, Starting Population**.

   17.1.   Answer question 6 on Discussion & Conclusions sheet.

18. Start the model running (click **GO**) and eat 25 snails by clicking the **EAT RANDOM SNAIL** button 25 times. Have the snails **REPRODUCE**.

19. To speed up the process of eating random snails, you can change the way the button works to eat more than one at a time. Click the number '1' to the immediate right of the **EAT RANDOM SNAIL** button, and select '25'. Then click the button one time, and you will instantly see 25 randomly selected snails disappear. **REPRODUCE**, have one more meal of 25 snails, and then **REPRODUCE** again.

20. Copy the histogram from the screen, paste it into your text document, and label it **Activity 2: No Selection, Generation 4—Trial 1**.

　20.1. **Answer question 7 on Discussion & Conclusions sheet.**

21. Reset the model. Then repeat steps 18-20, labeling the next histogram **Activity 2: No Selection, Generation 4—Trial 2**.

22. Reset the model and repeat steps 18–20 once more, labeling the histogram **Activity 2: No Selection, Generation 4—Trial 3**.

23. Look at all three Generation 4 histograms with no selection.

　23.1. **Answer question 8 & 9 on Discussion & Conclusions sheet.**

NOTE: In step 23, the mechanism of evolution is not natural selection. It is called *genetic drift*.

Lab 4 — Darwin's Theory of Evolution
ACTIVITY 3 — by Natural Selection

Here is how Charles Darwin thought adaptive evolution happens. Darwin said that:

A. if a population contains variation for some character, **AND**

B. if the variation is at least partly heritable (differences among individuals are at least partly due to differences in the genes they have inherited from their parents), **AND**

C. if some variants survive to reproduce at higher rates than others,

THEN the distribution of that character in the population will change over time.

Condition C, nonrandom survival and reproduction, is called *'natural selection'*. The individuals that survive to reproduce are said to be *naturally selected*. Together, the three conditions and the conclusion are **Darwin's Theory of Evolution by Natural Selection.**

PROCEDURE

Think about the results of the activities 1 and 2 and answer a question on Discussion & Conclusions sheet.

Lab 5
ISLE ROYALE

PPE-LAB READING

Textbook: Chapter 3: Popul... ...i Ecology.

Lab Manual: Lab 5

OBJECTIVES

- Understand the concepts of exponential growth, logistic growth, carrying capacity, simple food chains, and predator-pray dynamics.
- Use a simulation model to follow and describe the population sizes of moose and wolves as various parameters are changed within the simulation.
- Understand how model systems can be used in scientific research of population dynamics.
- Understand how models are being used to understand how ecosystems are likely to respond to the increased temperature or changes in precipitation resulting from the global climate change.
- Practice data interpretation in order to explain trends and observations.

MATERIAL

- SimBio Virtual Labs® EcoBeaker®: Isle Royale.

ACTIVITIES

1. *Review of Data Collection for Energy Lab*
2. The Moose Arrive
3. The Wolves Arrive
4. Changes in the Weather

Lab 5

ACTIVITY 1   Review of Data Collection for Energy Lab

PROCEDURE

Read Lab 9. Personal Energy Inventory: Activity 1—Data Collection. Discuss any questions about the data collection with your TA. The data have to be collected and recorded during the upcoming *Fall Break* week.

INTRODUCTION

▶ The Wolves and Moose of Isle Royale

If you were to travel on Route 61 to the farthest reaches of Minnesota and stand on the shore of Lake Superior looking east, on a clear day you would see Isle Royale. This remote, forested island sits isolated and uninhabited 15 miles off of the northern shore of Lake Superior, just south of the border between Canada and the USA. If you had been standing in a similar spot by the lake in the early 1900s, you may have witnessed a small group of hardy, pioneering moose swimming from the mainland across open water, eventually landing on the island. These fortunate
moose arrived to find a veritable paradise, devoid of predators and full of grass, shrubs, and trees to eat. Over the next 30 years, the moose population exploded, reaching several thousand individuals at its peak. The moose paradise didn't last for long, however.

Lake Superior rarely freezes. In the 1940s, however, conditions were cold and calm enough for an ice bridge to form between the mainland and Isle Royale. A small pack of wolves found the bridge and made the long trek across it to the island. Once on Isle Royale, the hungry wolves found their own paradise—a huge population of moose. The moose had eaten most of the available plant food, and many of them were severely undernourished. These slow-moving, starving moose were easy prey for wolves.

▶ The Isle Royale Natural Experiment

The study of moose and wolves on Isle Royale began in 1958 and is thought to be the longest-running study of its kind. The isolation of the island provides conditions for a unique natural experiment to study the predator-prey system. Isle Royale is large enough to support a wolf population, but small enough to allow scientists to keep track of all of the wolves and most of the moose on the island in any given year. Apart from occasionally eating beaver in the summer months, the wolves subsist entirely on a diet of moose. This relative lack of complicating factors on Isle Royale compared to the mainland has made the island a very useful study system for ecologists.

▶ The EcoBeaker® Version of Isle Royale

During this lab, you will perform your own experiments to study population dynamics using a computer simulation based on a simplified version of the Isle Royale community. The underlying model includes five species: *three plants* (grasses, maple trees, and balsam fir trees), *moose*, and *wolves*. If you were actually watching a large patch of moose-free grass through time, you would observe it slowly transforming into forest. Likewise, the simulated plant community exhibits a simple succession from grasses to trees.

While the animal species in the Isle Royale simulation are also simplified compared with their real-world counterparts, their most relevant behaviors are included in the model. Moose prefer to eat grass and fir trees. Wolves eat moose, more easily catching the slower, weaker moose. Each individual animal of both species has a store of fat reserves that decreases as the individual moves around and reproduces, and increases when food is consumed. Both moose and wolves reproduce; however, for simplicity, the simulation ignores gender. Any individual with enough energy simply duplicates itself, passing on a fraction of its energy to its offspring. Death occurs when an individual's energy level drops too low. Because weaker moose move at slower speeds, they take longer to find food and move away from predators, so their chance of survival is lower than for healthier moose. In the EcoBeaker simulation, wolves hunt alone, whereas in the real world, wolves are social animals that hunt in packs. These simplifications make the simulation tractable, while still retaining the basic qualitative nature of how these species interact.

SOME IMPORTANT TERMS AND CONCEPTS

▶ Population Ecology

Population ecology is the study of changes in the size and composition of populations and the factors that cause those changes.

▶ Population Growth

Many different factors influence how a population grows. Mathematical models of population growth provide helpful frameworks for understanding the complexity involved, and also (if the models are accurate) for predicting how populations will change through time. The simplest model of population growth considers a situation in which limitations to the population's growth do not exist (that is, all necessary resources for survival and reproduction are present in continual excess). Under these conditions, the larger a population becomes, the faster it will grow. If each successive generation has more offspring, the more individuals there will be to have even more offspring, and so on. This type of population growth is described with the exponential growth model.

The exponential growth model assumes that a population is increasing at its maximum per capita rate of growth (represented by 'r_{max}') also known as the "intrinsic rate of increase". If population size is N and time is t, then:

$$\frac{dN}{dt} = r_{max}N$$

The notation 'dN/dt' represents the "instantaneous change" in population size with respect to time. In this context, "instantaneous change" simply means how fast the population is growing or shrinking at any particular instant in time. The equation indicates that at larger values of N (the population size), the rate at which the population size increases will be greater.

The following graph depicts an example of exponential population growth. Notice how the curve starts out gradually moving upwards and then becomes steeper over time. This graph illustrates that when the population size is small, it can only increase in size slowly, but as it grows, it can increase more quickly.

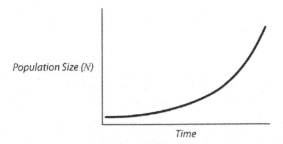

Exponential Population Growth

▶ Carrying Capacity

In the real world, conditions are generally not so favorable as those assumed for the exponential growth model. Population growth is normally limited by the availability of important resources such as food, nutrients, or space. A population's carrying capacity (symbolized by 'K') is the maximum number of individuals of that species that the local environment can support at any particular time. When a population is small, such as during the early stages of colonization, it may grow exponentially (or nearly so) as described above. As resources start to run out, however, population growth typically slows down and eventually the population size levels off at the population's carrying capacity.

To incorporate the influence of carrying capacity in projections of population growth rate, ecologists use the logistic growth model. In this model, the per capita growth rate (r) decreases as the population density increases. When the population is at its carrying capacity (i.e., when N = K) the population will no longer grow. Again, using the 'dN/dt' notation, if the maximum per capita rate of growth is rmax, population size is N, time is t, and carrying capacity is K, then:

$$\frac{dN}{dt} = r_{max}N\frac{(K - N)}{K}$$

When the population size (N) is near the carrying capacity (K), K – N will be small and hence, (K – N)/K will also be small. The change in the population size through time (dN/dt) will therefore decrease and approach zero (meaning the population size stops changing) as N gets closer to K.

The following graph depicts an example of logistic growth. Notice how it initially looks like the exponential growth graph but then levels off as N (population size) approaches K (carrying capacity).

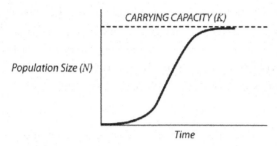

Logistic Population Growth

While the logistic model is more realistic than the exponential growth model for most populations, many other factors can also influence how populations change in size through time. For example, the growth curve for a recently-introduced species might temporarily overshoot the population's carrying capacity. This would happen if the abundance of resources encountered by the colonizing individuals stimulated a high rate of reproduction, but the pressures of limited resources were soon felt (i.e., individuals might not start dying off until after a period of rapid reproduction has already taken place).

Graphs based on real population data are never such smooth, neat curves as the ones above. Random events almost always cause population sizes and carrying capacities to fluctuate through time. Interactions with other species, such as predators, prey, or competitors, also cause the size of populations to change erratically. To estimate carrying capacity in situations such as these, one generally calculates the median value around which the population size is fluctuating.

▶ More Information

Links to additional terms and topics relevant to this laboratory can be found in the SimBio Virtual Labs Library which is accessible via the program's interface.

REFERENCES

A few researchers have studied the population dynamics of wolves and moose on Isle Royale for a very long time, resulting in an exceptional continuity in research approach and data collection. The research program is currently directed out of Michigan Tech by John Vucetich and Rolf Peterson, both of whom have published extensively on moose-wolf population dynamics. Below are a few references regarding moose and wolves on Isle Royale, the contribution of Isle Royale studies to broader ecological issues, and the scientific and conservation challenges involved.

Peterson, R.O., & Page, R.E. 1988. The Rise and Fall of Isle Royale wolves, 1975–1986. *Journal of Mammology,* 69: 89–99.

Peterson, R.O. 1995. *The Wolves of Isle Royale: A Broken Balance.* Willow Creek Press, Minocqua, WI.

Vucetich, J.A., R.O. Peterson, & C.L. Schaefer. 2002. The Effect of Prey and Predator Densities on Wolf Predation. *Ecology,* 83(11): 3003–3013.

Vucetich, J.A., & R.O. Peterson. 2004. Long-Term Population And Predation Dynamics Of Wolves On Isle Royale. In: D. Macdonald & C. Sillero-Zubiri (eds.), *Biology and Conservation of Wild Canids,* Oxford University Press, pp. 281–292.

Lab 5

Welcome to Isle Royale

PROCEDURE

1. Read the introductory sections of the workbook, which will help you understand what's going on in the simulation and answer questions.
2. Start **SimBio Virtual Labs®** by double-clicking the program icon on your computer or by selecting it from the Start Menu.
3. When the program opens, select the **Isle Royale** lab from the **EcoBeaker** suite.

   IMPORTANT!

   Before you continue, make sure you are using the **SimBio Virtual Labs** version of Isle Royale. The splash screen for SimBio Virtual Labs looks like this:

   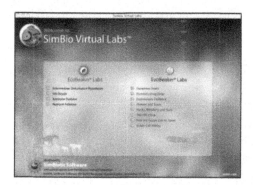

   If the splash screen you see does not look like this, please close the application (EcoBeaker 2.5) and launch **SimBio Virtual Labs**.

   When the **Isle Royale** lab opens, you will see several panels:

   The **ISLAND VIEW** panel (upper left section) shows a bird's eye view of northeastern Isle Royale, which hosts ideal moose habitat.

   The **DATA & GRAPHS** panel to the right displays a graph of population sizes of moose and wolves through time.

   The **SPECIES LEGEND** panel above the graph indicates the species in the simulation; the buttons link to the SimBio Virtual Labs Library where you can find more information about each.
4. Click 'Moose' in the **SPECIES LEGEND** panel to read about moose natural history, and then answer the following question (you should read about other species too).

   4.1. Based on what you find in the Library, answer the following: could a moose swim fast enough to win a swimming medal in the Olympics (where the fastest speeds are around 5 miles / hour)?

   Yes   No   (Circle one)

5. Examine the bottom row of buttons on your screen. You will use the **Control Panel** buttons to control the simulation and the **Tools** buttons (to the right) to conduct your experiments. These will be explained as you need them; if you become confused, position your mouse over an active button and a 'tool tip' will appear.

Lab 5

ACTIVITY 2 — The Moose Arrive

In this activity, you will study the moose on Isle Royale before the arrival of wolves. The lab simulates the arrival of the group of moose that swam to the island and rapidly reproduced to form a large population.

PROCEDURE

1. Click the **GO** button in the **Control Panel** at the bottom of the screen to begin the simulation. You will see the plants on Isle Royale starting to spread, slowly filling up most of this area of the island.

   Grass starts out as the most abundant plant species, but is soon replaced with maple and balsam fir trees. The Isle Royale simulation incorporates simplified vegetation succession to mimic the more complex succession of plant species that occurs in the real world. After about 5 simulated years, the first moose swim over to the island from the mainland and start munching voraciously on the plants.

2. You can zoom in or out using the **Zoom Level Selector** at the top of the **ISLAND VIEW** panel. Click different Zoom Level circles to view the action up closer or further away. After watching for a bit, click on the left circle to zoom back out. You can zoom in and out at any time.

3. Reset the simulation by clicking the **RESET** button in the **Control Panel**. Confirm that the simulation has been reset by checking that the **Time Elapsed** box to the right of the **Control Panel** reads "0 Years".

4. Click the **STEP 50** button on the **Control Panel**, and the simulation will run for 50 years and automatically stop. Watch the graph to confirm that the size of the moose population changes dramatically when the moose first arrive, and then eventually stabilizes (levels out).

   ★ *You can adjust how fast the simulation runs with the* **SPEED** *slider to the right of the* **CONTROL PANEL**.

5. Once 50 years have passed (model years—not real years!), examine the moose population graph and answer the questions below. (***NOTE:*** if you can't see the whole graph, use the scroll bar at the bottom of the graph panel to change the field of view.)

   5.1. Answer question 1-6 on the Data sheet.

6. Revisit the logistic growth equation in the Introduction and answer questions 1–7 on Discussion & Conclusions sheet.

8. Now you will test your prediction by increasing the number of moose on the island. Click the ADD MOOSE button in the Tools panel. With the ADD MOOSE button selected, move your mouse to the IsLAND VIEW, click and hold down the mouse to draw a small rectangle. As you draw, a number at the top of the rectangle tells you how many moose will be added. When you release the mouse, the new moose appear inside your rectangle. Add approximately 200-300 moose.

★   **HINT:** *To obtain the exact moose population size from the graph, click the graph to see the x and y data values at any point (population size is the y value).*

9. Click **GO** to continue running the simulation for 20 to 30 more years and watch what happens to the moose population. Click **STOP** to pause the simulation.

    9.1. Answer question 8–9 on Discussion & Conclusions sheet.

10. Click the **Test Your Understanding** button in the bottom right corner of the screen and answer the question in the window that pops up.

Lab 5
ACTIVITY 3
The Wolves Arrive

One especially cold and harsh winter in the late 1940s, Lake Superior froze between the mainland and Isle Royale. A small pack of wolves travelled across the ice from Canada and reached the island. In this activity, you will investigate how the presence of predators affects the moose population through time.

PROCEDURE

1. To load this activity, select "**The Wolves Arrive**" from the **Select an EXERCISE** menu at the top of the screen.
2. Click **STEP 50** to advance the simulation 50 years. You will see moose arrive and run around the island eating plants as before.

    2.1. Answer question 1 on Discussion & Conclusions sheet.

3. Add some wolves to the island. Activate the **ADD WOLF** button in the **Tools** panel by clicking it. Add 20–40 wolves to Isle Royale by drawing small rectangles on the island (they will fill with wolves) until you have succeeded in helping the wolf population to get established.
4. Run the simulation for about 200 years (you can click **STEP 50** four or five times). Observe how the moose and wolves interact, and how the population graph changes through time. (To better observe the system you can try changing the simulation speed or zoom level.)

    4.1. Answer questions 1–3 on the Data sheet.

5. If you haven't already, click **STOP**.
6. The **MICROSCOPE** tool lets you sample animals to determine their current energy reserves. Activate the **MICROSCOPE** tool by clicking it. Then click several moose to confirm that you can measure their 'Fat Stores'. These reserves are important health indicators for moose; the greater a moose's fat stores, the more likely it will survive the winter and produce healthy, viable offspring.

    6.1. Answer question 2 on Discussion & Conclusions sheet.

7. You will now test your prediction. **RESET** the simulation and then click **GO** to run the simulation without wolves until the moose population has stabilized at its carrying capacity. Click **STOP** so you can collect and record data. Decrease your zoom level to see as much of the island as possible.
8. Randomly select 10 adult moose and use the **MICROSCOPE** tool to sample their fat stores. Record your data on the left-hand side of the table below. Do NOT sample baby moose; they are still growing and so do not store fat as adults do.
9. When you are done, activate the **ADD WOLVES** button as before, and add 10-20 wolves. Click **GO** and run the simulation until the moose and wolf populations have cycled several times. **STOP** the simulation when the moose population is about midway between a low and high point (i.e. at its approximate average size).

10. Randomly select another 10 adult moose and use the **MICROSCOPE** tool to sample their fat stores.

    10.1. **Record the values on the right-hand side of the table 1 on Data sheet and answer question 5.**

11. Click the **Test Your Understanding** button and answer the question in the pop-up window.

Lab 5

ACTIVITY 4 — Changes in the Weather

You have probably heard that scientists are concerned about climate change and the effects of global warming due to increasing atmospheric greenhouse gases. Recent evidence suggests that temperatures around the world are rising. In particular, the average yearly temperature in northern temperate regions is expected to increase significantly. This change will lead to longer, warmer spring and summer seasons in places like Isle Royale. The duration of the growing season for plants will therefore be extended, resulting in more plant food for moose living on the island.

How would a longer growing season affect the moose and wolf populations on Isle Royale? Would they be relatively unaffected? Would the number of moose and wolves both increase indefinitely with higher and higher temperatures, and longer and longer growing seasons?

One way ecologists make predictions about the impacts of global warming is by testing different scenarios using computer models similar to the one you've been using in this lab. Even though simulation models are simplifications of the real world, they can be very useful for investigating how things might change in the future. In this activity, you will use the Isle Royale simulation to investigate how changes in average yearly temperature due to global warming may affect the plant-moose-wolf system on the island.

PROCEDURE

1. Use the **Select an EXERCISE** menu to launch "**Changes in the Weather**".
2. Click **STEP 50** to advance the simulation 50 years. You can zoom in to view the action up close. The moose population should level out before the simulation stops.
3. Activate the **ADD WOLF** button in the **TOOLS** panel. Add about 100 wolves by holding down your mouse button and drawing rectangular patches of wolves. Remember to look at the number at the top of the rectangle to determine how many wolves are added.
4. Advance the simulation 150 more years by clicking **STEP 50** three times. Watch the action. The simulation should stop at Year 200.

   4.1. Record the data in question 1 on Data sheet.

5. In the **PARAMETERS** panel below the **ISLAND VIEW** you will see "Duration of Growing Season" options where you can select different scenarios. The default is Normal, which serves as your baseline—this is the option you have been using thus far.

   ■ The Short option simulates a decrease in the average annual temperature on Isle Royale. The growing season is shorter than the baseline scenario, which results in annual plant productivity that is about half that of Normal.

■ The Long option simulates a warming scenario in which the growing season begins earlier in the spring and extends later in the autumn. Plant productivity is almost double that of Normal.

5.1. Answer question 1 on Discussion & Conclusions sheet.

6. Without resetting the model, select the '**Short**' growing season option.

7. Advance the simulation another 100 years by clicking **STEP 50** twice (total time elapsed should be ~300 years).

7.1. Record the data in question 2 and answer question 3 on the Data sheet.

8. In the short growing season, the plant growth is half of what it was before.

8.1. Answer question 4 on the Data sheet.

9. Now it's time to consider the warming scenario.

9.1. Answer question 2 on Discussion & Conclusions sheet.

10. Without resetting the model, select the '**Long**' growing season option from the **PARAMETERS** panel.

11. Click **GO** and monitor the graph as the populations cycle. If you watch for a while you should notice something dramatically different about this scenario, in which the plant productivity is high.

12. Click **STOP** and estimate the maximum and average size for moose and wolf populations under the Long growing season scenario.

12.1. Record the data in question 5 on the Data sheet and answer question 6.

13. Answer question 3 on Discussion & Conclusions sheet.

14. Click the **Test Your Understanding** button and answer the question in the pop-up window.

Lab 6
KEYSTONE PREDATOR

PPE-LAB READING

Textbook: Chapter 4: Species Interactions and Community Ecology.

Lab Manual: Lab 4.

OBJECTIVES

- Understand the concepts of community structure, trophic levels, food web, competition, and keystone species.
- Use a simulation model to determine relationships between different species in the community.
- Use a simulation model to understand a role of keystone species in a community.
- Understand how model systems can be used in scientific research of community processes and dynamics.
- Practice to interpret data in order to explain trends and observations.

MATERIAL

- SimBio Virtual Labs® EcoBeaker®: Keystone Predator.

ACTIVITIES

1. Flexing Your Mussels
2. You Are What You Eat
3. Who Rules the Rock?

INTRODUCTION

A diversity of strange-looking creatures makes their home in the tidal pools along the edge of rocky beaches. If you walk out on the rocks at low tide, you'll see a colorful variety of crusty, slimy, and squishy-looking organisms scuttling along and clinging to rock surfaces. Their inhabitants may not be as glamorous as the megafauna of the Serengeti or the bird life of Borneo, but these "**rocky intertidal**" areas turn out to be great places to study community ecology.

An **ecological community** is a group of species that live together and interact with each other. Some species eat others, some provide shelter for their neighbors, and some compete with each other for food and/or space. These relationships bind a community together and determine the local **community structure**: the composition and relative abundance of the different types of organisms present. The **intertidal community** is comprised of organisms living in the area covered by water at high tide and exposed to the air at low tide.

This laboratory is based on a series of famous experiments that were conducted in the 1960's along the rocky shore of Washington state, in the northwestern United States. Similar intertidal communities occur throughout the Pacific Northwest from Oregon to British Columbia in Canada. The nine species in this laboratory's simulated rocky intertidal area include three different algae (including one you may have eaten in a Japanese restaurant); three stationary (or "sessile") filter-feeders; and three mobile consumers.

Ecological communities are complicated, and the rocky intertidal community is no exception. Fortunately, carefully designed experiments can help us tease apart these complexities, providing insight into how communities function. As will become apparent, understanding the factors that govern community structure can have serious implications for management. In this laboratory, you'll use simulated experiments to elucidate how interactions between species can play a major role in determining community structure. You will apply techniques similar to those used in the original studies, in order to experimentally determine which species in the simulated rocky intertidal are competitively dominant over which others. You'll then analyze gut contents and use your data to construct a food web diagram. Finally, you'll conduct removal experiments, observing how the elimination of particular species influences the rest of the community. When you've completed this lab, you should have a greater appreciation for the underlying complexity of communities, and for how the loss of single species can have surprisingly profound impacts.

▶ Food Chains, Food Webs and Trophic Levels

You probably know that herbivores eat plants and that predators eat herbivores. The progression of what eats what, from plant to herbivore to predator, is an example of a **food chain**. Omnivores eat both plants and animals. Within a community, producers, herbivores, predators, and omnivores are linked through their feeding relationships. If you create a diagram that connects different species and food chains together based on these relationships, the result is called a **food web diagram**.

Ecosystems can also be represented by a pyramid comprising a series of "trophic levels". A species' trophic level indicates its relative position in the ecosystem's food chain. **Producers** (including algae and green plants) use energy from the sun to produce their own food rather than consuming other organisms, thus they occupy the lowest trophic level. Since herbivores consume the producers, they occupy the next trophic level. Predators eat the herbivores, thus occupying the next higher trophic level. Omnivores occupy multiple trophic levels. The highest level is occupied by top-level predators, which are not eaten by anything (until they die). Generally, but not always, lower trophic levels have more species than do higher levels within a community.

►Competition

Among community relationships, predation is perhaps the most obvious but certainly not the most important. Two species may also compete with each other for space or food. Stationary organisms in particular must often compete intensively for limited space. When one species is better at obtaining or holding space than another, or is able to displace the second species, the 'winner' is said to be **competitively dominant**. In the same way that you can draw a food web, you can also construct a diagram to illustrate which species are superior competitors within a community, called a **competitive dominance hierarchy**. In this lab, you will create competition dominancy hierarchy and food web diagrams to help you understand the community structure of the intertidal zone.

►Dominant versus Keystone Species

In many communities, there is one species that is more abundant in number or biomass than any other, often referred to as the **dominant species**. For example, in a dense, old-growth forest, one type of late successional tree is often the dominant species. As you might imagine, such species greatly affect the nature and composition of the community.

However, a species does not have to be the most abundant to have the greatest impact on the community. Imagine an archway made of stones. The one stone at the top center of the arch supports all the other stones. If you remove that stone, called the "keystone", the arch crumbles. In some communities, the presence of a single species controls community structure even though that species may have relatively low abundance. These organisms are known as **keystone species**. An important characteristic of a keystone species is that its decline or removal will drastically alter the structure of the local community. For example, many keystone species are top predators that keep the populations of lower-level consumers in check. If top predators are removed, populations of the lower-level organisms can grow, dramatically changing species diversity and overall community structure, sometimes resulting in the collapse of the entire community.

►The EcoBeaker® Model

If you're curious about how the simulated intertidal community works, here's the basic idea. In EcoBeaker models, each individual belongs to a "species" which is defined by a collection of rules that determine that species' behavior. For example, species that are mobile consumers follow rules that dictate how far they can move in a time step, what they can eat, how much energy they obtain from their prey, how much energy they use when they move, etc. When an individual consumer's energy runs out, it dies. Individuals within species all follow the same rules, but because the rules defining species include some random chance (e.g., which direction to turn), you will notice variability in what individuals are doing at any given time. Different species behave differently because they don't have the same "parameters" assigned for their rules (e.g., they might eat different species or move slower or faster).

The six stationary species in the model—the three algae and the three filter feeders (or "sessile consumers")—are modeled differently than the mobile consumers. The simulation uses a transition matrix for these six stationary species. The **transition matrix** is a set of probabilities that determine what happens from one time step to the next on a particular space on the rock. For each species, the transition matrix lists the probability of an individual of that species settling on top of bare rock, the probabilities of being replaced by each of the other species, and the probability of dying (and being replaced by bare rock). In addition, the transition matrix includes the probability of bare rock remaining bare. For example, a patch of rock that contains Nori Seaweed (*Porphyra*) may do one of three things each time step: host a different species (that is, another organism displaces Nori Seaweed), continue to be occupied by Nori Seaweed, or become bare rock (the Nori Seaweed dies and is not replaced). Each of these changes, or transitions, has a probability associated with it included within the transition matrix. If one species out-competes another for space, this will be reflected in the relevant transition probability.

▶ More Information

Links to additional terms and topics relevant to this laboratory can be found in the Keystone Predator Library accessible via the SimBio Virtual Labs® program interface.

REFERENCES

This laboratory was inspired by the following classic papers by R.T. Paine, whose work on intertidal communities originated the idea of keystone predation:

Paine, R.T. 1966. Food Web Complexity and Species Diversity. *The American Naturalist 100*: 65–75.
Paine, R.T. 1969. The Pisaster-Tegula Interaction: Prey Patches, Predator Food Preference, and Intertidal Community Structure. *Ecology* 50: 950–961.

The following review article addresses the idea of keystone species from a more modern perspective:

Power, M.E., D. Tilman, J.A. Estes, B.A. Menge, W.J. Bond, L.S. Mills, G. Daily, J.C. Castilla, J. Lubchenco, R.T. Paine. 1996. Challenges in the Quest For Keystones: Identifying Keystone Species is Difficult — But Essential To Understanding How Loss of Species Will Affect Ecosystems. *BioScience* 46: 609–620.

Lab 6

Welcome to the Rocky Intertidal Community

If you've explored tide pools (a fun thing to do if you visit a rocky coast), you likely know that many of the plants and animals living in them are unusual. This section will introduce you to the different species you'll encounter in this lab.

PROCEDURE

1. Make sure that you have read the introductory section of the workbook. The background information and introduction to ecological concepts will help you understand the simulation model and answer questions correctly.
2. Start the program by double-clicking the **SimBio Virtual Labs** icon on your computer or by selecting it from the **Start** menu.
3. When **SimBio Virtual Labs** opens, select the **Keystone Predator** lab from the **EcoBeaker** suite.

IMPORTANT!

Before you continue, make sure you are using the **SimBio Virtual** Labs version of Keystone Predator. The splash screen for SimBio Virtual Labs looks similar to this:

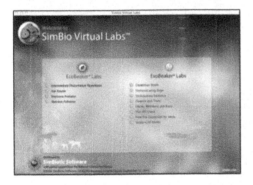

If the splash screen you see does not look like this, please close the application (EcoBeaker 2.5) and launch *SimBio Virtual Labs*.

You will see a number of different panels on the screen:

▪ The left side of the screen shows a view of an intertidal zone area. This is where you will be conducting your experiments and observing the action.
▪ A bar graph on the right shows the population sizes of all of species in the intertidal area.
▪ Above the graph is a list of the plant and animal species included in the simulation. You can switch between Latin and common names of each species using the tabs above the species list. The workbook will refer to common names.

▪ In the bottom left corner of the screen is the **Control Panel**. To the right of the **Control Panel** is a set of **TOOLS** that you will use for doing your experiments. These will be described in the following activities, as you need them.

4. Click on the names in the **Species Legend** in the upper right corner of the screen to bring up library pages for each species.

   4.1. Use the library to answer question 1 on Data sheet.

   4.2. Use the information in the Introduction and Library pages to fill in the blank spaces in a table on Data sheet.

5. When you are done completing the table, start the simulation by clicking the GO button in the **Control Panel**. Watch the action for a bit. Notice how the mobile consumers clear off areas of rock by eating, and how the stationary species recolonize those areas.

6. Click the **STOP** button to pause the simulation.

   Look at the population graph. The "**Population Size Index**" represents the number of individuals of each species present in the simulation. For the three algal species, a more appropriate measure of relative abundance might be percent cover or biomass, because in the real world, a single alga can grow quite large. The EcoBeaker model simulates algal growth as individuals multiplying, which, though not exactly realistic, makes possible the comparison of population sizes for the three algal and six animal species.

7. Click the **RESET** button to return the community to its initial state. Then move your mouse over to the population graph and click on one of the bars. You will see the population size for that species displayed.

   7.1. Answer questions 3-6 on the Data sheet.

Lab 6

ACTIVITY 2 You Are What You Eat

Now you will continue to investigate this intertidal community by determining what each of the mobile consumer species eats. One trick ecologists use to find out what creatures eat is to look at what's in their guts or excrement. If you eat something, it hangs around in your stomach for a little while, and then (later) the undigested parts come out the other end. Normally, when researchers look at gut contents they have to kill the animal, cut it open, and examine what is inside (people get paid to do this!). Within SimBio Virtual Labs there is a kinder and gentler method for determining gut contents.

PROCEDURE

1. Select "**You Are What You Eat**" from the **Select an Exercise** menu at the top of the screen.
2. You will now see all nine species in the simulation: three algae, three sessile consumers, and three mobile consumers: Starfish, Whelk, and Chiton.
3. Start running the simulation (click **GO**).
4. Watch the action for about 100 weeks and monitor the abundance of species in the population graph. Notice how the population index for each species fluctuates and eventually settles at a relatively stable level.
5. **STOP** the simulation.
6. Click the **MICROSCOPE** ("VIEW ORGANISM") button in the **Tools panel** to activate your mobile "Gut-o-Scope" (patent pending).
7. Click your favorite Starfish, Whelk, or Chiton—your choice!

   A window will appear with gut content information for that individual, either identifying the predator's last prey item or indicating that the gut is empty (because the creature has not eaten recently). Note that if you click on organisms that don't have guts (algae or filter feeders), you won't see gut contents.
8. You will now conduct a survey of the three mobile consumers to learn which species they eat.

   8.1 Use table 1 on the Data sheet to record and summarize your findings. There are three table sections, one for each mobile consumer.

   8.2 Fill up table 2 on the Data sheet.

9. You now have enough information to construct a food web diagram from your findings. Consider the hypothetical example below. In a forest, both deer and rabbits eat the plants. Wolves, the predators in the system, eat both deer and rabbits. We can draw these feeding relationships like this, with arrows pointing **to** the consumer:

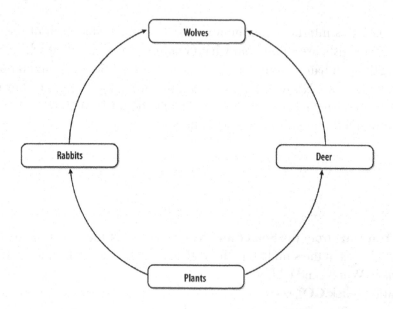

9.1. Construct a diagram on the Discussion & Conclusions sheet.

10. Click the **Test Your Understanding** button and answer the question in the pop-up window.

Lab 6

ACTIVITY 2 — You Are What You Eat

Student Name: Nina Zhang   Lab Group: B

TA: Amanda Tuminelli   Lab Date/Section: _____

DATA

1. Table 1. In each data table section, record gut content data for 10 randomly-selected individuals of that species. Ignore individuals that have not eaten recently (indicated by gut contents labeled "empty"). If recorded correctly, each row of the data form should have one species circled. At the bottom of each section, record the total number of individuals circled in each column.

GUT CONTENT DATA FOR WHELK

1.	Nori	Black Pine	Coral Weed	Mussel	(Acorn)	Gooseneck	Whelk	Chiton	Starfish
2.	Nori	Black Pine	Coral Weed	Mussel	(Acorn)	Gooseneck	Whelk	Chiton	Starfish
3.	Nori	Black Pine	Coral Weed	Mussel	(Acorn)	Gooseneck	Whelk	Chiton	Starfish
4.	Nori	Black Pine	Coral Weed	Mussel	(Acorn)	Gooseneck	Whelk	Chiton	Starfish
5.	Nori	Black Pine	Coral Weed	Mussel	Acorn	(Gooseneck)	Whelk	Chiton	Starfish
6.	Nori	Black Pine	Coral Weed	Mussel	(Acorn)	Gooseneck	Whelk	Chiton	Starfish
7.	Nori	Black Pine	Coral Weed	Mussel	(Acorn)	Gooseneck	Whelk	Chiton	Starfish
8.	Nori	Black Pine	Coral Weed	Mussel	Acorn	(Gooseneck)	Whelk	Chiton	Starfish
9.	Nori	Black Pine	Coral Weed	Mussel	(Acorn)	Gooseneck	Whelk	Chiton	Starfish
10.	Nori	Black Pine	Coral Weed	Mussel	(Acorn)	Gooseneck	Whelk	Chiton	Starfish
	#___	#___	#___	#___	#8	#2	#___	#___	#___

GUT CONTENT DATA FOR CHITON

1.	Nori	Black Pine	(Coral Weed)	Mussel	Acorn	Gooseneck	Whelk	Chiton	Starfish
2.	Nori	Black Pine	(Coral Weed)	Mussel	Acorn	Gooseneck	Whelk	Chiton	Starfish
3.	(Nori)	Black Pine	Coral Weed	Mussel	Acorn	Gooseneck	Whelk	Chiton	Starfish
4.	Nori	(Black Pine)	Coral Weed	Mussel	Acorn	Gooseneck	Whelk	Chiton	Starfish
5.	Nori	Black Pine	(Coral Weed)	Mussel	Acorn	Gooseneck	Whelk	Chiton	Starfish
6.	Nori	Black Pine	(Coral Weed)	Mussel	Acorn	Gooseneck	Whelk	Chiton	Starfish
7.	Nori	Black Pine	(Coral Weed)	Mussel	Acorn	Gooseneck	Whelk	Chiton	Starfish
8.	Nori	Black Pine	(Coral Weed)	Mussel	Acorn	Gooseneck	Whelk	Chiton	Starfish
9.	Nori	Black Pine	(Coral Weed)	Mussel	Acorn	Gooseneck	Whelk	Chiton	Starfish
10.	(Nori)	Black Pine	Coral Weed	Mussel	Acorn	Gooseneck	Whelk	Chiton	Starfish
	#2	#1	#7	#___	#___	#___	#___	#___	#___

GUT CONTENT DATA FOR STARFISH

1.	Nori	Black Pine	Coral Weed	Mussel	Acorn	Gooseneck	Whelk	Chiton	Starfish
2.	Nori	Black Pine	Coral Weed	Mussel	Acorn	Gooseneck	Whelk	Chiton	Starfish
3.	Nori	Black Pine	Coral Weed	Mussel	Acorn	Gooseneck	Whelk	Chiton	Starfish
4.	Nori	Black Pine	Coral Weed	Mussel	Acorn	Gooseneck	Whelk	Chiton	Starfish
5.	Nori	Black Pine	Coral Weed	Mussel	Acorn	Gooseneck	Whelk	Chiton	Starfish
6.	Nori	Black Pine	Coral Weed	Mussel	Acorn	Gooseneck	Whelk	Chiton	Starfish
7.	Nori	Black Pine	Coral Weed	Mussel	Acorn	Gooseneck	Whelk	Chiton	Starfish
8.	Nori	Black Pine	Coral Weed	Mussel	Acorn	Gooseneck	Whelk	Chiton	Starfish
9.	Nori	Black Pine	Coral Weed	Mussel	Acorn	Gooseneck	Whelk	Chiton	Starfish
10.	Nori	Black Pine	Coral Weed	Mussel	Acorn	Gooseneck	Whelk	Chiton	Starfish
	#_____	#_____	#_____	#_____	#_____	#_____	#_____	#_____	#_____

2. Table 2. For each mobile consumer species below, record its prey and the percentage of diet each prey species comprises for that consumer (e.g., 4 out of 10 samples = 40%). The numbers in each column should add up to 100%.

	Whelk	Chiton	Starfish
Prey Species:			
Percentage of Diet:			
Prey Species:			
Percentage of Diet:			
Prey Species:			
Percentage of Diet:			
Prey Species:			
Percentage of Diet:			

Lab 6

ACTIVITY 2 You Are What You Eat

Student Name: Nina Zhang   Lab Group: B

TA: Amanda Tuminelli   Lab Date/Section: _____

DISCUSSION & CONCLUSIONS

Use your data on feeding relationships to construct a food web diagram for the organisms that live in the simulated intertidal zone. Link the species names below with arrows that point from prey to consumer. (Unlike the simple four-species example in the text, your nine-species diagram will look more complicated, with many crossing lines.)

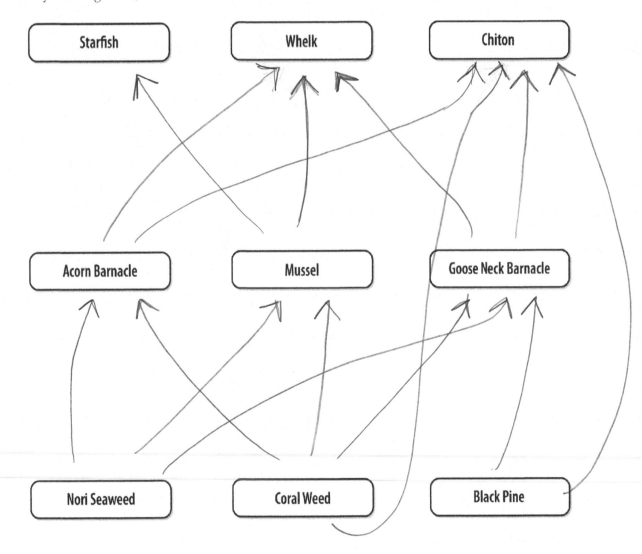

Lab 6

ACTIVITY 3 — Who Rules the Rock?

Based on your studies so far, you know important details about the competitive and feeding relationships among the species in your simulated intertidal community, and these relationships, when integrated, define the role played by each. One way to more fully elucidate the importance of a species to its community structure is to remove it from the environment and observe what happens. In this activity, you will experimentally determine how removing each of the highest trophic level species (the mobile consumers) affects the rocky intertidal community structure.

PROCEDURE

1. Before you start your experiments, you will first make some predictions. Refer back to your data to inform your answers. [NOTE: only one species will be removed in each experiment.]

   1.1. **Answer questions 1 and 2 on the Discussion & Conclusions sheet.**

2. Select "**Who Rules the Rock?**" from the **Select an Exercise** menu.

3. Your first step is to record population sizes BEFORE REMOVALS. To make sure the simulation is initialized correctly, click the **RESET** button. A data table is provided on the next page for recording your results.

   HELFUL HINT: *if you click on the colored bars in* **the Population Size** *graph, the numbers (population sizes) that the bars represent will pop up!*

   3.1. **In the table 1 on the Data sheet, record the population size of each species at 'Time Elapsed = 0 Weeks' in the BEFORE REMOVALS column.**

4. After recording data BEFORE REMOVALS, you are ready to remove mobile consumers. Find the **REMOVE WHELK** button (which depicts a Whelk with a slash through it) in the **Tools Panel**. When you click this button, all Whelk will vanish from the Intertidal Zone.

5. For each removal experiment, you will run the simulation for 200 weeks (in model time, not real time!). To do this, first make sure that the Time Elapsed = 0 weeks (RESET if not), and then click the **STEP 200** button in the **control panel**.

   HELFUL HINT: *If your computer is a little slow, you can speed things up using the Speed Slider to the right of the* **control panel**

6. Confirm that the simulation stopped at (or near) 200 weeks. If so, click the bars in the Population Size graph and record the abundance of each species.

   6.1. **In the data table 1, record the population size of each species in the AFTER WHELK REMOVAL column.**

7. **Reset** the simulation and confirm that Time Elapsed = 0 weeks. Then click the **REMOVE chiton** button to remove all Chiton from the Intertidal Zone.

8. Click the **STEP 200** button to run the simulation for 200 weeks.

   8.1. When Time Elapsed = 200 weeks, record the population size of each species in the AFTER CHITON REMOVAL column in the table 1.

9. Finally, **RESET** the simulation and use the **remove starfish** tool and the step 200 button to repeat the experiment for Starfish.

   9.1. In the data table 1, record the population size of each species in the AFTER STARFISH REMOVAL column.

10. When your data table 1 is complete, answer questions 3-9 on the Discussion & Conclusions sheet.

11. Click the **Test Your Understanding** button and answer the question in the pop-up window.

Lab 7

SURVIVORSHIP IN POPULATIONS

PRE-LAB READING

Textbook: Chapter 8: Human Population.

Lab Manual: Lab 6.

ACTIVITIES

1. Human Data Collection
2. Comparison of Curves

INTRODUCTION

Different types of organisms exhibit different types of survivorship. **Survivorship** is an expression of the proportion of individuals that survive an established interval of the life span.

Survivorship curves are a graphic representation of survivorship data for a population. The y-axis shows the number of individuals. The x-axis can show specific age categories of time intervals, but frequently use the percent of an organism's potential life span. By using percent of potential life span, one can graph the survivorship curves of different types of organisms on the same plot for comparison.

There are three general types of survivorship curves; Type I, II, and III. These curves are illustrated in Figure 7.1 below.

Organisms that exhibit a **Type I** survivorship curve have little mortality in the early part of the life span. The majority of individuals survive to each reproductive age. The greatest amount of mortality occurs in the later portion of the life span. Examples of organisms that exhibit a Type I survivorship curve are humans, elephants, and polar bears. These types of organisms have a greater probability of dying later in life than early in life.

Organisms that exhibit a **Type III** survivorship curve experience the majority of their mortality in the early part of the potential life span. A small percentage of the initial number of individuals survive to reproductive age but those that do survive that long tend to go on to survive for the majority of the remaining potential life span. Examples of organisms that exhibit a Type III survivorship curve include many insects and annual plants. These types of organisms have a greater probability of dying early in life than later in life.

Organisms that exhibit a **Type II** survivorship curve exhibit a pattern that is intermediate relative to the other two types of curves. For these organisms, the probability of dying is basically the same at any time during their life span.

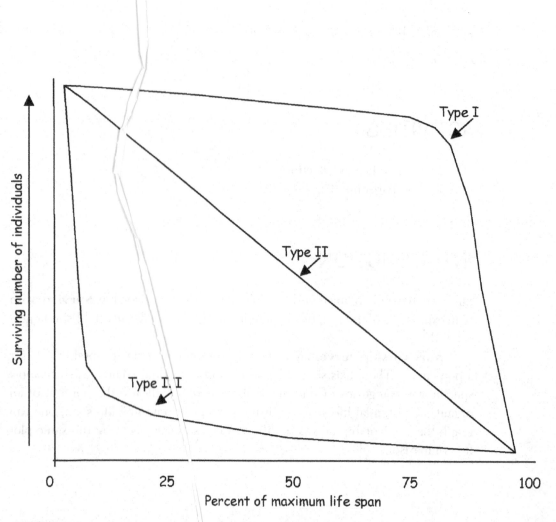

FIGURE 7.1.

Approximate shapes of Type I, II, and III survivorship curves as number of individuals versus percent of potential life span.

In this exercise, you will compile human survivorship data from on-line records and construct a survivorship curve using the data. You will also prepare survivorship curves using survivorship data for several other species and compare those curves to the human curve.

REFERENCES

Anonymous. (n.d.) Population dynamics problem set. Retrieved May 16, 2010. From http://www.tulane.edu/~ggentry/ECOL/NuSet2004.doc.

Anonymous. 2007. Oak Grove Cemetery, Delaware, Delaware County, Ohio. *Internment.net.* Retrieved July 28, 2007. From http://www.interment.net/data/us/oh/delaware/oakgrove/.

Begon, Michael, and M. M. Mortimer. 1981. Population ecology: a unified study of animals and plants. Blackwell Scientific Publications, Oxford, England.

Buikema, Arthur. 2004. Survivorship Curves. Retrieved July 28, 2007. From http://www.bioinquiry.vt.edu/bioinquiry/Cheetah/cheetahpaid/cheetahhtmls/popsurvivor.htmlhttp://www.bioinquiry.vt.edu/bioinquiry/Cheetah/cheetahpaid/cheetahhtmls/popsurvivor.html.

Bush, Mark B. 2003. Ecology of a changing planet, 3rd edition. Prentice Hall, Upper Saddle River, New Jersey.

Horne, J. S. 2009. WLF 448 fish and wildlife population ecology lab notes. *Fish and Wildlife Population Ecology.* Retrieved May 16, 2010. From http://www.cnr.uidaho.edu/wlf448/2008/Lab/9a_Problems.htm.

Kevan, P. G., and O. Kukal. 1993. Corrigendum: A balanced life table for *Gynaephora groenlandica* (Lepidoptera: Lymantriidae), a long-lived high arctic insect, and implications for the stability of its populations. Can. J. Zool. 71:1699–1701. Retrieved May 16, 2010. From http://article.pubs.nrc-cnrc.gc.ca/ppv/RPViewDoc?issn=1480-3283&volume=71&issue=8&startPage=1699.

Rockwood, L. 2003. Biology 307: General ecology lab manual. George Mason University, Fairfax, Virginia.

Lab 7

ACTIVITY 1 Human Data Collection

OBJECTIVES

- ■ Understand the concepts related to age-specific survivorship in populations.
- ■ Understand the type of survivorship curve exhibited by human populations.
- ■ Construct a survivorship curve for a human population.

HYPOTHESIS

- ■ A human population that lived in the 1800s will exhibit a Type I survivorship curve.

MATERIALS

- ■ Computer access to on-line cemetery records, containing information on a population from a localized area with a sufficient number of records for individuals that lived during the 1800s.

PROCEDURE

▶ **Part A: Data Collection**

1. Each lab group will begin by working as two **pairs**, and later will get back together as their lab group of four.
2. Work in pairs to collect data:
   Accessing data on-line: Your lab class will use the computers in the lab classroom to access cemetery data from http://www.interment.net/data/us/oh/delaware/oakgrove/. Each pair will collect data for the surnames ranges indicated in Table 7.1.

TABLE 7.1. Surname Ranges Assigned to Each Pair for Human Data Collection from On-Line Cemetery Records

LAB GROUP #1		LAB GROUP #2	
Pair 1	Pair 2	Pair 1	Pair 2
A	Bo-Bri	Cl-Co	Di-Dy
Ba	Bro-By	Cr-Cz	E
Be-Bl	Ca-Ci	Da-De	Fa-Fl
LAB GROUP #3		**LAB GROUP #4**	
Pair 1	Pair 2	Pair 1	Pair 2
Fo-Fy	He-Hn	Ka-Ki	Mc-Me
Ga-Go	Ho-Hy	Kl-Ky	Mi
Gr-Gw	I	La-Le	Mo-My
Ha	J	Li-Ly	N
		Ma	O
LAB GROUP #5		**LAB GROUP #6**	
Pair 1	Pair 2	Pair 1	Pair 2
Pa-Pf	Ro-Ry	St-Sz	Wa
Ph-Py	Sa-Se	Ta-Th	We-Wh
Q	Sh-Sk	Ti-Ty	Wi
Ra-Ri	Sl-Sr	U	Wo-Wy
		V	X-Z

3. "Sample" (collect data about) each deceased individual that meets the following criterion:
   a. Birth year was **between** 1800 and 1899, inclusive
4. For each sampled individual, determine and record in Table 7.2 on the data sheet their **gender** and **age at death**.
   a. The determination of gender must sometimes be based on information such as the first name of the individual, references to occupation, or use of terms such as "loving father," "loving mother," etc. If you cannot determine the gender, do not sample the individual.
   b. Table 7.2 provides space to record birth year and death year but the only information needed is **age at death**. Some students prefer to record birth and death year while sampling and then calculate and record age at death back in the lab. Other students will "do the math" while sampling and just record the age at death.
5. Continue working as a pair to gather data for Table 7.2 until you either
   a. finish your assigned section of the on-line records (at which time you should ask your instructor if any other sections need to be surveyed) OR
   b. fill up all available spaces on Table 7.2

▶ Part B: Data Compilation

1. Student pairs will work together as a **lab group** to combine and compile their data into Table 7.3 on the data sheet, as follows:
   a. Review the Table 7.2 data collected by each student pair.
   b. The column on Table 7.3 labeled "tally" is just an area that you can use to place tick marks as you compile the data by each student pair into a single set of group data.

    c. Determine the total number of males that died in each age category (as defined in the first column of Table 7.3) and enter that group-based data into the appropriate rows in Table 7.2.

    d. Repeat step c for females.

2. Record your group's data final data from Table 7.3 in the appropriate columns and rows of Table 7.4 on the data sheet.

3. Record in Table 7.4 in the appropriate columns the data from the other lab groups (this data might have been written on a black/white board, overhead transparency, or computer projected image).

4. Complete Table 7.4 by calculating the total number of deaths in each age category by gender across all groups.

5. Using the class-wide data from Table 7.4, calculate and record in Table 7.5 on the data sheet the number of surviving individuals by gender and age class and the total number of surviving individuals by age class.

    For example, the number of surviving males for the age category "<1" would be the total number of males in all age class minus the total number of males that died in the "<1" age category.

Lab 7
ACTIVITY 1
Human Data Collection

Student Name: _____   Lab Group:_____

TA: _____   Lab Date/Section: _____

DATA

TABLE 7.2. Age at Death Data for Individuals by Gender, Collected by a Student Pair

MALES			FEMALES		
Birth Year	Death Year	Age at Death	Birth Year	Death Year	Age at Death

Lab 7

ACTIVITY 1 Human Data Collection

Student Name: _____   Lab Group:_____

TA: _____   Lab Date/Section: _____

DATA

TABLE 7.3. Student-Pair Data for Age at Death by Gender Compiled into Lab
          Group Data for Age at Death by Age Class and Gender

Age Class	MALE DEATHS		FEMALE DEATHS	
	Tally	Total	Tally	Total
< 1				
1–4				
5–9				
10–14				
15–19				
20–24				
25–29				
30–34				
35–39				
40–44				
45–49				
50–54				
55–59				
60–64				
65–69				
70–74				
75–79				
80–84				
85–89				
90–94				
95–99				
100–104				
105–109				
110+				

Lab 7

ACTIVITY 1

Human Data Collection

Student Name: _____ Lab Group:_____

TA: _____ Lab Date/Section: _____

/

DATA

TABLE 7.4. Age at Death Data by Gender and Age Class for All Groups

Age Class	Male Deaths (by lab group #)							Female Deaths (by lab group #)						
1	2	3	4	5	6	Total	1	2	3	4	5	6	Total
< 1													
1–4													
5–9													
10–14													
15–19													
20–24													
25–29													
30–34													
35–39													
40–44													
45–49													
50–54													
55–59													
60–64													
65–69													
70–74													
75–79													
80–84													
85–89													
90–94													
95–99													
100–104													
105–109													
110+													

Lab 7
ACTIVITY 1
Human Data Collection

Student Name: _____ Lab Group:_____

TA: _____ Lab Date/Section: _____

DATA

TABLE 7.5. Number Deaths and Surviving Individuals by Age Class for Males, Females, and Males and Females Combined for Class-Wide Data

Age Class	MALES		FEMALES		MALES + FEMALES	
	# Deaths	# Surviving	# Deaths	# Surviving	# Deaths	# Surviving
< 1						
1–4						
5–9						
10–14						
15–19						
20–24						
25–29						
30–34						
35–39						
40–44						
45–49						
50–54						
55–59						
60–64						
65–69						
70–74						
75–79						
80–84						
85–89						
90–94						
95–99						
100–104						
105–109						
110+						

Lab 7

ACTIVITY 1   Human Data Collection

Student Name: _____   Lab Group:_____

TA: _____   Lab Date/Section: _____

DISCUSSION & CONCLUSIONS

For full credit, questions should be answered thoroughly, in complete sentences, and legibly.

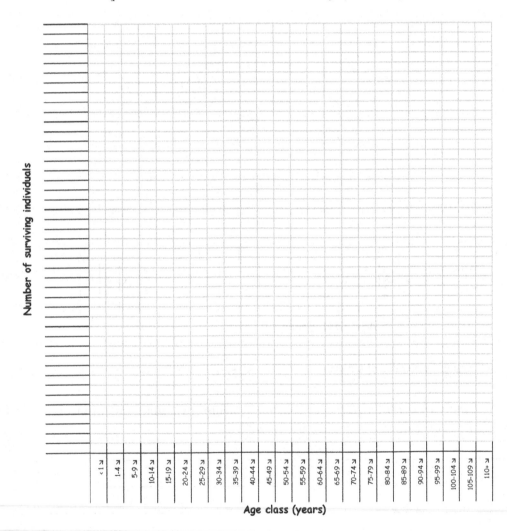

FIGURE 7.2.
Human survivorship as number of survivors by age class (years).

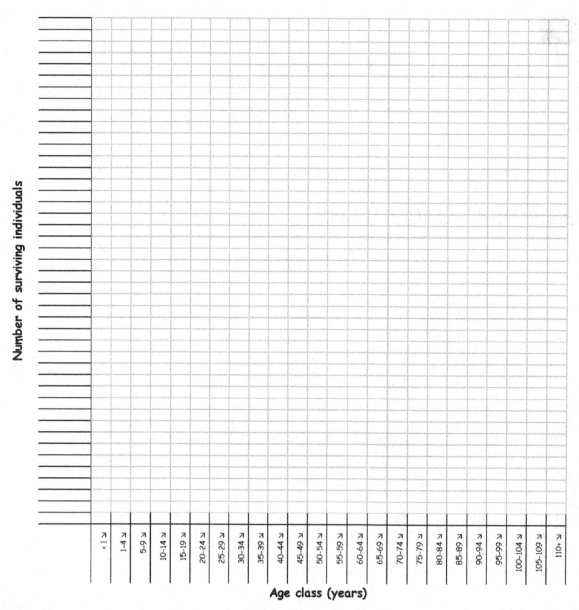

FIGURE 7.3.
Human male and female survivorship as number of survivors by age class (years).

1. *Which type survivorship curve (I, II, or III) does* your survivorship curve in Figure 6.1 most closely resemble? Is this *consistent* with your expectations? *Why?*

_____

_____

_____

_____

_____

_____

2. Based on your Figure 7.2, *which* age class seem to have the highest levels of mortality (lowest survivorship)? *Do* males and females exhibit their highest levels of mortality in the same age intervals? Can you suggest *reasons* for this?

_____

_____

_____

_____

_____

_____

_____

_____

3. *Identify* possible sources of error in the execution of this activity.

_____

_____

_____

_____

For each hypothesis listed below, state whether or not it was supported by the data collected and explain your response.

4. *Hypothesis:* A human population that lived in the 1800s will exhibit a Type I survivorship curve.

_____

_____

_____

_____

_____

_____

_____

_____

_____

_____

_____

_____

Lab 7

ACTIVITY 2 Comparison of Curves

OBJECTIVES

▪ Understand the concepts related to age-specific survivorship in populations.
▪ Construct survivorship curves using survivorship data for a variety of species.

HYPOTHESIS

▪ Different species will exhibit different survivorship curves.

MATERIALS

▪ Calculator
▪ Human data from activity 1.

PROCEDURE

▶ Part A: Data for Humans

1. Refer to your human data from the last column of Table 7.5 in Activity 1—Human Data Collection. That column provides the number of surviving individuals by age class for males and females combined. You are going to "re-compile" that data to make it easier to compare to data for other species.
2. Calculate and record in Table 7.6 the *total* number of humans (males and females combined) surviving by the new age classes indicated in the table. Note that age class "0" represents the original total population.
3. Calculate and record in Table 7.6 the *percent* of humans (males and females combined) surviving by the new age classes indicated in the table. Note that age class "0" represents the original total population and would be 100%.

▶ Part B: Data for Other Species

1. Tables 7.7–7.11 contain data on the number of surviving by age class for the Arctic high moth, bull trout, ground squirrel, annual bluegrass, and finch, respectively.
2. Calculate and record in the last column of Tables 7.7–7.11 the percent of individuals surviving by age class.

Lab 7
ACTIVITY 2
Comparison of Curves

Student Name: Nina Zhang

Lab Group: B

TA: Amanda Tuminelli

Lab Date/Section: _____

DATA

$\dfrac{\text{\# surviving}}{\text{total pop}}$ $\dfrac{231}{241}$

TABLE 7.6. Human Survivorship Data as Number and Percent Surviving by Age Class for Class-Wide Data from Activity 1

Age Class #	Age Range for Class (yrs)	Human Survivorship	
		# Surviving	% Surviving
0	Initial pop. size	241	100
1	0-9	231	95.8
2	10-19	228	94.6
3	20-29	219	90.8
4	30-39	210	87.1
5	40-49	196	80.9
6	50-59	174	73.4
7	60-69	147	60.9
8	70-79	79	32.7
9	80-89	10	4.141
10	90-99	5	2.070
11	≥100	2	0.8282

TABLE 7.7. High Arctic Moth Survivorship Data as Number and Percent Surviving by Age Class

Age Class #	Age Range for Class	High Arctic Moth Survivorship	
		# Surviving	% Surviving
0	Egg (= Initial pop. size)	85	100
1	Instar I	71	83.5
2	Instar II	59	69.4
3	Instar III	49	57.6
4	Instar IV	26	30.5
5	Instar V	14	16.4
6	Instar VI	7	8.23
7	Pupa	2	2.35
8	Adult moth	1	1.17

Lab 7

ACTIVITY 2

Comparison of Curves

Student Name: Nina Zhang   Lab Group: B

TA: Amanda T   Lab Date/Section: _____

DATA

TABLE 7.8. Bull Trout Survivorship Data as Number and Percent Surviving by Age Class

Age Class #	Age Range for Class (yrs)	Bull Trout Survivorship	
		# Surviving	% Surviving
0	Initial pop. size	3000	100
1	0–1	200	6.66
2	1–2	150	5
3	2–3	75	2.5
4	3–4	70	2.3
5	4–5	45	1.5
6	5–6	31	1.03
7	6–7	14	0.46
8	7–8	10	0.33
9	8–9	6	0.2

TABLE 7.9. Ground Squirrel Survivorship Data as Number and Percent Surviving by Age Class for Class-Wide Data from Activity 1

Age Class #	Age Range for Class	Ground Squirrel Survivorship	
		# Surviving	% Surviving
0	Initial pop. size	686	100
1	0–1	500	72.8
2	1–2	235	34.25
3	2–3	101	14.72
4	3–4	46	6.705
5	4–5	21	3.06
6	5–6	9	1.311
7	6–7	5	0.728
8	7–8	4	0.583
9	8–9	1	0.145

Lab 7

ACTIVITY 2 Comparison of Curves

Student Name: Nina Zhang   Lab Group: B

TA: Amanda T   Lab Date/Section: _____

DATA

TABLE 7.10. Annual Bluegrass Survivorship Data as Number and Percent Surviving by Age Class

Age Class #	Age Range for Class (months)	Annual Bluegrass Survivorship	
		#Surviving	% Surviving
0	Initial pop. size	1000	100
1	0–3	842	84.2
2	3–6	640	64
3	6–9	366	36.6
4	9–12	190	19
5	12–15	59	5.9
6	15–18	19	1.9
7	18–21	3	0.3
8	21–24	0	0

TABLE 7.11. Finch Survivorship Data as Number and Percent Surviving by Age Class

Age Class	Age Range for Class (yrs)	Finch Survivorship	
		# Surviving	% Surviving
0	Initial pop. size	284	100
1	0–1	158	55.6
2	1–2	139	48.9
3	2–3	117	41.1
4	3–4	109	38.3
5	4–5	87	30.63
6	5–6	63	22.1
7	6–7	35	12.3
8	7–8	28	9.85
9	8–9	15	5.28

Lab 8

SOIL CHARACTERIZATION

PRE-LAB READING

Textbook: Chapter 9: Soils.

Lab Manual: Lab 7.

ACTIVITIES

1. Soil Texture
2. Soil pH and Nutrient Content

INTRODUCTION

Soil is an important component of the physical environment. It provides a substrate in which plants grow and from which they obtain required mineral nutrients. It is also the medium through which plants obtain their water, taking it in at their roots. The soil provides a home for a large variety of microorganisms that perform a large variety of vital ecosystem services, including nitrogen fixation, decomposition, and nutrient cycling.

All ecosystems have at their base producers—organisms that can make their own food using inorganic compounds and an energy source. In terrestrial ecosystems the predominant producers are plants. Even though plants are autotrophic, they require water and mineral nutrients in order to function and survive and plants obtain these materials via the soil.

The importance of plants to humans cannot be overstated. They supply all of our food, directly or indirectly (for the most part, we either eat plants or eat animals that ate plants!). They affect climate, provide us with oxygen, and provide structural materials and fiber. Many humans value the aesthetic contributions of plants.

The character of the soil in a particular area affects what type plants, and, therefore, what type of ecosystems, can develop and be supported in that area. Some of the many parameters of soil character are its texture, permeability, porosity, pH, and nutrient content (such as nitrogen and phosphorus).

Before discussing the parameters of soil character, it is important to understand what soil is.

Soil is a mixture of small particles (<2 mm diameter) of minerals that have been weathered from parent rock material by physical and chemical means, and organic matter.

The inorganic component of soil is mineral particles that fall into three classes based on their diameter (according to USDA-adopted scheme); **sand** (0.05–2.0 mm in diameter), **silt** (0.002–0.05 mm in diameter), and **clay** (<0.002 mm in diameter). Particles of inorganic material with diameters greater than 2 mm are called **gravel** and are not technically part of the soil.

To determine the texture of a soil, it is necessary to first separate and remove the gravel portion. Then, the relative percentages of sand, silt, and clay particles are determined. The soil texture category is determined using the information on the percentages of sand, silt, and clay particles, and a soil texture diagram.

Soil texture affects plant growth in many ways including influencing the ability of plant roots to move through the soil, the amount of oxygen in the soil, and the water-holding capacity of the soil.

Soil porosity is a measure of the pore space in soil. It is physical measure that does not reflect the ease with which water will pass through the soil. **Soil permeability** is a measure of the ease with which water will pass through the soil. It is a function of both the physical aspect of particle and pore size and electrochemical properties that influence how tightly water is attracted to and held by and between the particles.

Soil pH is a measure of the acidity or alkalinity of the soil. It is influenced by soil type and the pH of rain in the area. It has an effect on all types of life that might inhabit the soil.

The presence or absence in soil of nutrients essential for plant growth affects what types of plants can grow in an area. Nitrogen, phosphorus, and potassium are three nutrients essential for plant growth.

REFERENCES

Anonymous. (n.d.). What is Soil? *U.S. Department of Agriculture, Natural Resources Conservation Service.* Retrieved July 28, 2007. From http://soils.usda.gov/education/facts/soil.html.

Brown, R. B. 2003. Soil texture. Fact Sheet SL-29, Soil and Water Science Department, Florida Cooperative Extension Service, Institute of Food and Agricultural Sciences, University of Florida. Retrieved 11 April 2010. From http://edis.ifas.ufl.edu/pdffiles/SS/SS16900.pdf.

Padilla, M. J., M. Cyr, J. D. Exline, I. Miaoulis, J. M. Pasachoff, B. B. Simons, C. G. Vogel, and T. R. Wellnitz. 2001. Science explorer: earth science. Prentice- Hall, Inc., Upper Saddle River, New Jersey.

Withgott, Jay and S. Brennan. 2008. Environment: the science behind the stories, 3rd edition. Pearson Education Inc., California.

Lab 8

ACTIVITY 1 Soil Texture

OBJECTIVES

- Learn a technique for determining soil texture.
- Determine the % sand, % clay, and % silt of soil samples from two sources.
- Determine soil texture category using data on % sand, % clay, and % silt of soil samples from two sources and information in a soil texture diagram.
- Understand that soil texture will vary for soil from different sources.

HYPOTHESES

- There will be a difference in the soil texture of soil collected from difference sources.
- The forest soil samples collected by different lab groups will not have the same soil texture.
- The construction soil samples collected by different lab groups will not have the same soil texture.

MATERIALS

- Dustpan and brush, small, 1
- Filter, coffee, basket-type, 1
- Mortar and pestle, 1
- Sieve, 2 mm mesh, 1
- Soil samples, 2 sources
- Soil texture test kit, at least 1, preferably 2
- Tray, 1

PROCEDURE

1. Empty ~1/2 of the soil from the "forest" bag into the mortar and use the pestle to composite (mix thoroughly, removing coarse root material) the soil into a homogeneous mixture.
2. Pass the composited soil sample through a 2 mm sieve, allowing the material that passes through the mesh to fall onto a tray (or butcher paper or newspaper, rather than onto the lab table).

3. The portion of the soil sample that stays in the top of the sieve is the **gravel** (because its particles are >2 mm). The gravel is not considered soil. You can discard the **gravel**, as directed by your instructor.

4. Use a dustpan and brush to sweep the soil up off the tray (or paper) and place it into a coffee filter (on which you have written the source of this soil sample), which will contain the soil until you are ready to set up the soil texture test. The material that *did* pass through the sieve is considered **soil**, consisting of particles of **sand**, **silt** and **clay**, and will be used to determine soil texture.

5. Repeat steps 1–4 for the soil sample from the "construction" location.

6. You now have two coffee filters, containing the sieved portions of your samples.

7. You have been provided with at least one soil texture test kit. If so, you will have to process each of your two samples one at a time. If you have been provided with two kits, you can process your two samples concurrently.

8. Follow the instructions in the soil texture analysis kit to determine the % clay, % sand, and % silt for both of your soils. Record these results in Table 8.1 on the data sheet.

9. Use the soil texture diagram in Figure 8.1 on the next page to determine the soil texture category of each soil sample, based on the data collected in step #8 above. To use the soil texture diagram:

    a. Find the % sand value of your sample on the sand side of the triangle. Sketch a line (either mentally, or in light pencil so that it can later be erased, or with a different color for each soil sample) from that point across the triangle such that the line is parallel to the side of the triangle that is opposite of the 100% sand corner of the triangle.

    b. Find the % silt value of your sample on the silt side of the triangle. "Sketch" (as explained above) a line from that point across the triangle such that the line is parallel to the side of the triangle that is opposite of the 100% silt corner of the triangle.

    c. Find the % clay value of your sample on the silt side of the triangle. "Sketch" (as explained above) a line from that point across the triangle such that the line is parallel to the side of the triangle that is opposite of the 100% clay corner of the triangle.

    d. Theoretically, the three lines will intersect at a single point within the soil texture diagram. In reality, this might not happen perfectly because of your estimation of the location of the smaller values on the diagram. In this case, your lines might only come close to intersecting. The labeled area of the soil texture diagram within which the lines intersect (or come close to intersecting) is the soil texture category for the soil sample.

    e. For example, the soil texture diagram would indicate the "sandy clay loam" texture category for a soil sample that was 72% sand, 3% silt, and 25% clay.

10. Record the soil texture category for each soil sample in Table 8.1 on the data sheet.

11. Also record the soil texture category for each soil sample in Table 8.3 on the data sheet associated with Activity 2: Soil pH and Nutrient Content."

12. **CLEANUP:** When you are finished, make sure the following cleanup steps have been carried out:

    a. Rinse and dry the tray (or discard the butcher paper).

    b. Use the dustpan and brush to sweep up loose soil.

    c. Clean the lab table with wet paper towels.

    d. DO NOT DISCARD ANY SOIL IN THE SINKS!!!!!

    e. Empty remaining soil from soil texture test kit tubes into the trash cans.

    f. Clean and return all soil texture kit components to the kit(s) on your lab table.

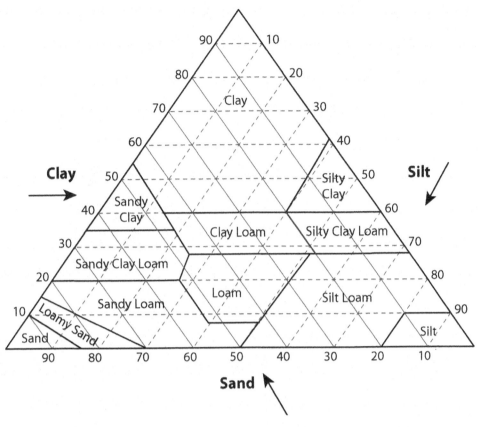

FIGURE 8.1.
Soil texture diagram for use in determining soil texture category based on the % clay, % silt, and % sand composition of the soil sample. (Diagram source: U.S. Department of Agriculture).

Lab 8

ACTIVITY 1   Soil Texture

Student Name: _____   Lab Group:_____

TA: _____   Lab Date/Section: _____

DATA

TABLE 8.1. Percent Sand, Silt and Clay, and Soil Texture Category for Two Soil Samples from Different Sources

Soil Source	% Sand	% Silt	% Clay	Texture Category
1.				
2.				

Lab 8

ACTIVITY 1   Soil Texture

Student Name: _____   Lab Group:_____

TA: _____   Lab Date/Section: _____

DISCUSSION & CONCLUSIONS

For full credit, questions should be answered thoroughly, in complete sentences, and legibly.

1. *Identify* possible sources of error in the execution of this activity.

_____

_____

_____

_____

For the hypothesis listed below, state whether or not it was supported by the data collected and explain your response.

Hypothesis: There will be a difference in the soil texture of a forest soil sample versus a construction soil sample.

_____

_____

_____

_____

_____

_____

_____

_____

_____

_____

_____

_____

Lab 8
ACTIVITY 2
Soil pH and Nutrient Content

OBJECTIVES

■ Learn techniques for determining soil pH, and nitrogen, phosphorus, and potassium content.
■ Determine the pH, and nitrogen, phosphorus, and potassium content of soil samples from two sources.
■ Understand that soil pH, and nitrogen, phosphorus and potassium content will vary in soil samples from different sources.

HYPOTHESES

■ There will be a difference in the soil pH of two soils from different sources.
■ There will be a difference in the soil nitrogen content of two soils from different sources.
■ There will be a difference in the soil phosphorus content of two soils from different sources.
■ There will be a difference in the soil potassium content of two soils from different sources.

MATERIALS

■ Dustpan and brush, small, 1
■ Filter, fine pore, ~20 cm diameter, 4
■ Flask, Erlenmeyer, 125 mL
■ Funnel, ~10 cm diameter mouth, ~2 cm diameter tip, 2
■ Liquid extract from forest and construction soil samples (prepared in activity 1)
■ Nitrogen test kit, 1
■ pH test kit, 1
■ Phosphorus test kit, 1
■ Potassium test kit, 1

PROCEDURE

1. Use the two flasks from activity 1 for the tests in this activity.
2. If the liquid in the flasks is murky, re-filter through a fine pore filter using the filter, funnel, and Erlenmeyer flask. Repeat the filtering process a third time if needed.
3. You have been provided with pH, nitrogen, phosphorus, and potassium test kits (or test strips). Follow the instructions in kits (or on strip containers) to carry out all four tests on both of your soil samples. Record the results of the tests in Table 8.2 on the data sheet. Also record these results in Table 8.3.
4. Record in Table 8.3 on the data sheet your group's soil texture results from activity 1.
5. Complete Table 8.3 by recording in the appropriate columns the data from the other lab groups (this data might have been written on a black/white board or computer-projected image).
6. **CLEANUP:** When you are finished, make sure the following cleanup steps have been carried out:
   a. Discard the filter(s), if used, in the trash.
   b. Pour any remaining liquid extract down the sink.
   c. Wash and dry the 125 mL flasks, making sure to wipe off the wax markings.
   d. Return all test kits components to the kit(s) on your lab table.

Lab 8
ACTIVITY 2
Soil pH and Nutrient Content

Student Name: _____   Lab Group:_____

TA: _____   Lab Date/Section: _____

DATA

TABLE 8.2. Soil pH, and Nitrogen, Phosphorus (as nitrate and phosphate), and Potassium Content for Soil from Two Sources, Tested by Individual Lab Group

Soil Source	pH	Nitrate (units: _____)	Phosphate (units: _____)	Potassium (units: _____)
1				
2				

TABLE 8.3. Summary of All Soil Characterization Data for Soil from Two Sources for All Groups

Character / Soil Source → / Group # ↓	Texture Category[1] 1.	2.	pH 1.	2.	Nitrate (__) 1.	2.	Phosphate (__) 1.	2.	Potassium (__) 1.	2.
1										
2										
3										
4										
5										
6										

[1] Texture category abbreviations:
C = Clay   S = Sand   Si = Silt
CL = Clay Loam   SaC = Sandy Clay   SiC = Silty Clay
L = Loam   SaCL = Sandy Clay Loam   SiCL = Silty Clay Loam
LS = Loamy Sand   SaL = Sandy Loam   SiL = Silt Loam

Lab 8

ACTIVITY 2

Soil pH and Nutrient Content

Student Name: _____   Lab Group:_____

TA: _____   Lab Date/Section: _____

DISCUSSION & CONCLUSIONS

For full credit, questions should be answered thoroughly, in complete sentences, and legibly.

1. *Identify* possible sources of error in the determining the pH of the soil samples.

_____

_____

_____

2. *Identify* possible sources of error in the determining the nutrient content of the soil samples.

_____

_____

_____

3. *Why* are soil pH and nutrient content important to ecosystems?

_____

_____

_____

For each hypothesis listed below, state whether or not it was supported by the data collected and explain your response.

4. *Hypothesis:* There will be a difference in the soil pH of two soils from different sources.

_____

_____

_____

_____

5. *Hypothesis:* There will be a difference in the soil nitrogen content of two soils from different sources.

_____

_____

_____

_____

6. *Hypothesis:* There will be a difference in the soil phosphorus content of two soils from different sources.

_____

_____

_____

_____

7. *Hypothesis:* There will be a difference in the soil potassium content of two soils from different sources.

_____

_____

_____

_____

Lab 9

PERSONAL ENERGY INVENTORY

PRE-LAB READING

Textbook: Chapter 19: Sources of Energy. Energy Efficiency and Conservation.

Lab Manual: Lab 9.

ACTIVITIES

1. Data Collection
2. Data Compilation

INTRODUCTION

All human activities require use of energy, directly or indirectly. Since the time of industrial revolution, fossil fuels (oil, coal, and natural gas) are our society dominant source of energy. Global consumption of fossil fuels has risen steadily over the past half-century and now at its highest level ever. But not all people and all countries in the world have the same fuel and energy needs and use level. People in the developed countries with the higher quality of life usually consume disproportionate share of the world's energy. For example, the United States has only 4.5% of the world's population, but it consumes over 20% of the world's energy. Each year the average U.S. citizen uses 17 times more energy that does the average citizen of India.

In Activity 1, you will conduct a personal energy inventory of the approximate amount of energy you use in one regular day. You will calculate how much of that energy use is for transportation and how much is for non-transportation purposes. Please use Table 9.1 for your inventory when you are not able to determine

TABLE 9.1. Watt Usage Assumptions for Various Household Appliances

Appliance	Typical Wattage per Hour
Air Conditioner (12,000 BTU)	1500
Air Conditioner (36,000 BTU)	4500
Blender	385
CD, Tape, Radio, Receiver System	250
Clock	3
Clothes Dryer (high)	5000
Clothes Dryer (med)	3000
Clothes Washer	512
Coffee Maker (Auto Drip)	1165
Computer (With Monitor and Printer)	365
Convection Oven	1500
Curling Iron	1500
Dishwasher (Dry Cycle)	1200
Dishwasher (Wash Cycle)	200
Disposal	420
Fan (Ceiling)	80
Freezer (Automatic Defrost 15 cu. ft.)	440
Freezer (Manual Defrost, 15 cu. ft.)	350
Garage Door Opener	350
Hair Dryer (Hand Held)	1000
Heater (Portable)	1500
Heating System (Warm Air Fan)	312
Humidifier (Winter)	177
Iron	1000
Jacuzzi (Maintain Temperature, 2 Person)	1500
Lighting (Incandescent)	75
Lighting (Fluorescent)	40
Lighting (Compact Fluorescent)	18
Microwave Oven	1500
Mixer, Hand	100
Power Tools (Circular Saw)	1800
Radio	71
Range (high, 1 burner)	5000
Range (med, 1 burner)	2500
Range (Oven @ 350°F)	2660
Range (Self Cleaning Cycle)	2500
Refrigerator/Freezer (Frostfree, 17.5 cu.ft.)	450
Satellite Dish (Includes Receiver)	360
Television (Color, Solid State)	200
Toaster	1400
Vacuum Cleaner	1560
VCR/DVD	21

TABLE 9.2. Transportation Vehicle Fuel Mileage Assumptions

Vehicle	Fuel mileage
Car	25 mpg
Mini-van, SUV, PU	18 mpg
Bus	10 mpg

the actual watt usage of a device or appliance. Table 9.2 provides mileage assumptions that can be used for determining the amount of energy you use for transportation.

In Activity 2, you will compare your daily transportation and non-transportation energy use to that of your classmates. You will conduct a statistical test to determine if there is a difference between the transportation energy use among students who live on campus versus those who commute everyday. In addition, you will also conduct a statistical test to determine is there is a difference between the non-transportation energy use among the students who live alone versus those that live in group.

REFERENCES

Brennan, S., and J. Withgott. 2011. Environment: The science behind the stories, 4th edition. Pearson Education, San Francisco, California.

"Where Does All That Electricity Go?" 2002. Operating Costs of Household Appliances. Nebraska Public Power District. Retrieved June 16, 2011. From http://www.cornhusker-power.com/householdappliances.asp

World Resources Institute. 2000. World resources 200-2001. World Resources Institute, Washington, D.C.

Lab 9
ACTIVITY 1   Data Collection

OBJECTIVES

- Become aware of one's daily level of personal energy use.
- Inventory one's own personal energy use for one day.
- Calculate daily personal energy use for non-transportation versus transportation purposes.

HYPOTHESIS

- None.

MATERIALS

- Data sheet
- Observation skills

PROCEDURE

1. Work individually,
2. Chose a day, between now and your **NEXT** lab class to monitor your energy usage by category. You should select a day that is representative of your overall lifestyle.
3. *For one day,* monitor and record your energy use in three categories; (1) non-transportation, without hot water demand, (2) non-transportation, with hot water demand, and (3) transportation.
4. **Non-transportation, without hot water demand energy use** will be recorded in Table 9.3 of the data sheet. Use this table to record non-transportation uses that DO NOT involve hot water,
   a. Column A is the category name.
      i. Multiple rows have been provided for incandescent and fluorescent light bulbs since your usage will probably include bulbs of more than one type and wattage.
   b. Column B is for the watts used by the item.
      i. If possible, try to record the actual watts used by an item (look for a label on the device with this information).

    ii. If you cannot determine the actual watts used by a specific item, refer to Table I.2 (in the introduction section of this exercise), which provides typical watts used for various devices.

    c. Column C is for kilowatts, which you will **calculate** by dividing the watts (column B) by 1000.

    d. Column D is a measure of usage in minutes (you should track your usage in minutes).

    e. Column E is for the number of hours the item was used, which you will calculate by dividing the minutes used (column D) by 60.

    f. Column F is for kilowatt hours used, which you will calculate by multiplying the kilowatts (column C) by the hours used (column E).

    g. Total (sum up) all the values for kilowatt hours used in each category (column F) and record this number in the last row of the table as your total non-transportation, without hot water demand, energy usage.

5. **Non-transportation, *with* hot water demand energy use** will be recorded in Table 9.4 of the data sheet. Use this table to record non-transportation uses that DO involve hot water, such as showers, baths, dishwashing, and clothes washing. PLEASE NOTE that for this category you must determine the minutes and/or the number of times that the device was used as well as determining the quantity of water used.

    a. Column A is the category name and there are two main subcategories;

        i. **Electricity used by the appliance:**

            1) Column B is for the watts used by the item.

                a) If possible, try to record the actual watts used by an item (look for a label on the device with this information).

                b) If you cannot determine the actual watts used by a specific item, refer to Table 9.1 (from introduction of this exercise) which provides typical watts used for various household appliances.

            2) Column C is for kilowatts, which you will **calculate** by dividing the watts (column B) by 1000.

            3) Column D is a measure of usage in minutes (you should track your usage in minutes).

            4) Column E is for the number of hours the item was used, which you will **calculate** by dividing the minutes used (column D) by 60.

            5) Column F is for kilowatt hours used, which you will **calculate** by multiplying the kilowatts (column C) by the hours used (column E).

        ii. **Electricity/energy used to heat water:**

            1) For dish washing and clothes washing:

                a) **Record** in column B the number of loads.

                b) **Calculate** the number of gallons of hot water used by multiplying column B times the value provided in column C and record the result in column D.

                c) **Calculate** the kilowatt hours used by multiplying the # gallons used (column D) by the conversion factor provided in column E and record the result in column F.

            2) For showering:

                a) **Record** in column A the length of the shower in minutes.

                b) **Calculate** the number of gallons of hot water used by ' multiplying column B times the value provided in column C and record the result in column D.

                c) **Calculate** the kilowatt hours used by multiplying the # gallons used (column D) by the conversion factor provided in column E and record the result in column F.

3) For bathing:

    a) Record in column B the number of baths taken.

    b) Calculate the number of gallons of hot water used by multiplying column B times the value provided in column C and record the result in column D.

    c) Calculate the Kilowatt hours used by multiplying the # gallons used (column D) by the conversion factor provided in column E and record the result in column F.

  b. Total (sum up) the values for kilowatt hours used (column F) and record this number in the last row of the table as your total non-transportation, with hot water demand, energy usage.

6. **Transportation energy use** will be recorded in Table 9.5 of the data sheet.

  a. Record in column A the total miles traveled.

    i. Record miles traveled regardless of whether those miles were traveled by your own car, in a friend's car, in a taxi, or on a bus,

    ii. If multiple people were in the travel vehicle (as in carpooling, sharing a taxi, taking a bus), divide the miles by the number of people in the vehicle.

  b. Record in column B the fuel mileage (in miles per gallon) of the vehicle.

    i. If you don't know the vehicle's fuel mileage, select an approximate figure from Table 9.2 (in the introduction section of this exercise).

  c. Calculate the gallons of fuel used by dividing the miles traveled (column A) by the miles per gallon (column B) and record the result in column C.

  d. Calculate the transportation energy use in megajoules by multiplying the total gallons used (column C) by 10 (because each gallon of gas contains energy equivalent to 10 mega joules) and record the result in the last row of the table as the total energy use in mega joules.

7. Convert the **non-transportation** energy use to megajoules.

  a. Transfer the total non-transportation energy use, without hot water demand, in kilowatt hours from the last row and column of Table 9.3 of the data sheet to Table 9.6.

  b. Transfer the total non-transportation energy use, *with* hot water demand, in kilowatt hours from the last row and column of Table 9.4 of the data sheet to Table 9.6.

  c. Calculate the energy used in Megajoules by multiplying kilowatt hours (column B) times the conversion factor provided in column C and record the result in column D.

  d. Complete Table 9.6 of the data sheet adding the values in column D and recording the result in the last row of the table as the total non-transportation energy use in mega joules.

8. In Table 9.7 of the data sheet record the transportation (from Table 9.5) and non-transportation (from Table 9.6) energy use totals in mega joules.

  a. Add the values and record the result in the last row of Table 9.7 of the data sheet as your *total daily* energy use in mega joules.

NOTE: There is no discussion and conclusions section for this activity.

Lab 9

ACTIVITY 1 Data Collection

Student Name: _____ Lab Group:_____

TA: _____ Lab Date/Section: _____

DATA

TABLE 9.3. Personal Daily Non-Transportation Energy Use for Activities without Hot Water Demand, by Category, in Kilowatt-Hours

A	B	C	D	E	F
Device Name/Category	Watts (from device label or chart)	Kilowatts (Watts/1000)	Minutes Used (count)	Hours Used (minutes/60)	Kilowatt Hours Used (C×E)
Incandescent bulb					
Incandescent bulb					
Incandescent bulb					
Fluorescent bulb					
Fluorescent bulb					
Fluorescent bulb					
Refrigerator					
Electric stove					
Microwave oven					
Television					
VCR					
DVD player					
Stereo w/CD player					
MP3 player					
Radio					
Computer					
Clothes dryer					
Coffee maker					
Clock					
Non-Transportation Energy Use, WITHOUT Hot Water Demand				TOTAL:	

Lab 9

ACTIVITY 1 — Data Collection

Student Name: _____   Lab Group:_____

TA: _____   Lab Date/Section: _____

DATA

TABLE 9.4. Personal Daily Non-Transportation Energy Use for Activities with Hot Water Demand, by Category, in Kilowatt-Hours

→

A	B	C	D	E	F
Electricity to run device: → Device name/category: ↓	Watts (from device label or chart)	Kilowatts (B/1000)	Minutes Used (count)	Hours Used (b/60)	Kilowatt Hours Used (C×E)
Dishwasher					
Clothes Washer					
Electricity to heat water: → Device name/category: ↓	# Loads (count)	Gallons/Load	# Gallons (B×C)	Conversion Factor	Kilowatt Hours Used (D×E)
Dishwasher		15		0.195	
Clothes Washer		5		0.195	
Electricity to heat water: → Device name/category: ↓	# Minutes (count)	Gallons/Min	# Gallons (B×C)	Conversion Factor	Kilowatt Hours Used (D×E)
Showers		2		0.195	
Electricity to heat water: → Device name/category: ↓	# Taken (count)	Gallons/Bath	# Gallons (B×C)	Conversion Factor	Kilowatt Hours Used (D×E)
Baths		20		0.195	
Non-Transportation Energy Use, WITH Hot Water Demand				TOTAL:	

Lab 9

ACTIVITY 1 — Data Collection

Student Name: _____   Lab Group:_____

TA: _____   Lab Date/Section: _____

DATA

TABLE 9.5. Personal Daily Transportation Energy Use in Megajoules

	A	B	C	D
Category:	Miles Traveled	Miles/Gal	Gallons Used (A/B)	Megajoules (C×10)
Travel				
Transportation Energy Use			TOTAL:	

TABLE 9.6. Personal Daily Energy Use, All Non-Transportation Categories, in Megajoules

A	B	C	D
Category:	Kilowatt-hours	Conversion Factor	Megajoules (B×C)
Non-transportation w/out hot water demands		3.6	
Non-transportation w/hot water demands		3.6	
Non-transportation Energy Use		TOTAL:	

TABLE 9.7. Personal Daily Energy Use, Non-Transportation Versus Transportation, in Megajoules

Energy Use Category	Megajoules
Non-transportation	
Transportation	
Energy Use TOTAL:	

Lab 9
ACTIVITY 2 Data Compilation

OBJECTIVES

▪ Become aware of how one's daily level of personal energy use compares with that of other students.
▪ Compare daily personal energy use for non-transportation versus transportation purposes.
▪ Compare daily personal energy use for non-transportation purposes with classmates.
▪ Compare daily personal energy use for transportation purposes with classmates.
▪ Compare daily personal energy use for non-transportation purposes between students who live alone versus those who live in groups.
▪ Compare daily personal energy use for transportation purposes between students who live on-campus versus those who live off-campus.

HYPOTHESES

▪ Personal energy use will be greater for non-transportation purposes than for transportation purposes.
▪ There will be no difference in personal energy use for non-transportation purposes for students who live in groups than for students who live alone.
▪ There will be no difference in personal energy use for transportation purposes for students who live off-campus than for students who live on-campus.
▪ Personal energy use will vary greatly among students due to differences in personal habits, living arrangements, modes of transportation, and diligence in carrying out the exercise.

MATERIALS

▪ Personal energy inventory data from activity 1

PROCEDURES

1. Work individually.
2. Record your name, your daily non-transportation and transportation energy use totals in mega joules (from Table 9.7, activity 1), whether or not you live on-campus or off-campus, and whether

or not you live alone or in a group on Table 9.8 on the data sheet (and on the blackboard, transparency, or computer).

3. Copy from the transparency (or blackboard) all classmates' data to your Table 9.8 on the data sheet.

4. Transfer to Table 9.9 on the data sheet the transportation energy use data from Table 9.8 for students based on living location. Note that you do not need the student names any longer. Complete the table by calculating the mean daily transportation energy use for students living on-campus versus those living off-campus.

5. To determine if the difference between the means of transportation energy use for students living on-campus versus off-campus is significant, that is to say that the difference is not due to chance, use the statistical test called the student's t-test.

6. Go to http://www.physics.csbsju.edu/ stats/t-test_bulk_form.html. This website provides a simple on-line tool for calculating the student's t-test of two means.

7. Once at the site, you will see two boxes labeled "data for group A" and "data for group B."

8. Into the box labeled "data for group A" type each of the values from the first column of Table 9.9 on the data sheet. Separate the numbers with a comma but no spaces.

9. Into the box labeled "data for group B" type each of the values from the second column of Table 9.9 on the data sheet. Separate the numbers with a comma but no spaces.

10. After you have typed the data into both boxes, click on "calculate now."

11. A new page, headed "student's t-test: results," will appear. From this page, copy into Table 9.10 on the data sheet the value of t and the probability value.

12. If the probability value is less than or equal to 0.05, then the difference between the mean transportation energy use for students living on-campus versus student's living off-campus is considered "statistically significant." That is to say that the difference cannot be attributed to chance alone. You would then reject your hypothesis (reject that there is no difference). If the probability value is greater than 0.05, then the difference is not considered statistically significant and you would then accept your hypothesis (that there is no difference).

13. Indicate in Table 9.10 on the data sheet whether your hypothesis is rejected or accepted.

14. Complete Table 9.10 on the data sheet by formulating and recording a conclusion statement, such as "daily transportation energy use for students living on-campus is greater than for students living off-campus."

15. Transfer to Table 9.11 on the data sheet the **non-transportation** energy use data from Table 9.11 for students based on living status. Note that you do not need the student names any longer. Complete the table by calculating the mean daily non-transportation energy use for students living alone versus those living in a group.

16. To determine if the difference between the means of transportation energy use for students living on-campus versus off-campus is significant, that is to say that the difference is not due to chance, use the statistical test called the student's t-test.

17. Go to http://www.physics.csbsju.edul stats/t-test_bulk_form.html to conduct a student's t-test of the daily mean non-transportation energy use for students living alone versus those living in a group.

18. Into the box labeled "data for group A" type each of the values from the first column of Table 9.11 on the data sheet. Separate the numbers with a comma but no spaces.

19. Into the box labeled "data for group B" type each of the values from the second column of Table 9.11 on the data sheet. Separate the numbers with a comma but no spaces.

20. After you have typed the data into both boxes, click on "calculate now."

21. A new page, headed "student's t-test: results," will appear. From this page, copy into Table 9.12 on the data sheet the value of t and the probability value.

22. If the probability value is less than or equal to 0.05, then the difference between the mean non-transportation energy use for students living alone versus students living in a group is considered "statistically significant." That is to say that the difference cannot be attributed to chance alone. You would then reject your hypothesis (reject that there is no difference). If the probability value is greater than 0.05, then the difference is not considered statistically significant and you would then accept you hypothesis (that there is no difference).

23. Complete Table 9.12 on the data sheet by formulating and recording a conclusion statement, such as "daily non-transportation energy use for students living alone is less than for students living in a group."

Lab 9
ACTIVITY 2
Data Compilation

Student Name: _____   Lab Group:_____

TA: _____   Lab Date/Section: _____

DATA

TABLE 9.8. Daily Non-Transportation and Transportation Energy Use (Mergajoules), Living
Location (On- Or Off-Campus), and Living Status (Alone or in a Group), by Student

Student Name	Living Location		Living Status		Energy Use (Megajoules)	
	On-Campus	Off Campus	Alone	In Group	Non-Transportation	Transportation
					Mean =	

Lab 9

ACTIVITY 2 Data Compilation

Student Name: _____   Lab Group:_____

TA: _____   Lab Date/Section: _____

DATA

TABLE 9.9. Mean Daily Transportation Energy Use (Megajoules) for Students Living Off-Campus Versus Students Living On-Campus

Student Daily Transportation Energy Use (megajoules) by Living Location	
On-Campus	Off-Campus
Mean:	Mean:

TABLE 9.10. Results of Student's t-Test of Difference Between Means of Daily Transportation Energy Use (Megajoules) for Students Living On-Campus Versus Students Living Off-Campus

T value:		Probability:	
Null Hypothesis Rejected or Accepted:			
Conclusion:			

Lab 9
ACTIVITY 2   Data Compilation

Student Name: _____   Lab Group:_____

TA: _____   Lab Date/Section: _____

DATA

TABLE 9.11. Mean Daily Non-Transportation Energy Use (Megajoules) for
Students Living Alone Versus Students Living in a Group

Student Daily NON-Transportation Energy Use (mega joules) by Living Statis	
Alone	In a Group
Mean:	Mean:

TABLE 9.12. Results of Student's t-Test of Difference Between Means of Daily Non-Transportation
Energy Use (Megajoules) for Students Living Alone Versus Students Living in a Group

T value:		Probability:	
Null Hypothesis Rejected or Accepted:			
Conclusion:			

Lab 9

ACTIVITY 2

Data Compilation

Student Name: _____   Lab Group:_____

TA: _____   Lab Date/Section: _____

DISCUSSION & CONCLUSIONS

For full credit, questions should be answered thoroughly, in complete sentences, and legibly.

1. *What* are some steps you could take to reduce your non-transportation energy use?

_____

_____

_____

_____

2. *What* are some steps you could take to reduce your transportation energy use?

_____

_____

_____

_____

For each hypothesis below, state whether or not it was supported by the data collected and explain your response.

3. *Hypothesis:* Personal energy use will be greater for non-transportation purposes than for transportation purposes.

_____

_____

_____

_____

_____

_____

_____

4. *Hypothesis:* There will be no difference in persona/ energy use for non-transportation purposes for students who live in groups than for students who live alone.

_____

_____

_____

_____

_____

_____

_____

_____

5. *Hypothesis:* There will be no difference in personal energy use for transportation purposes for students who live off-campus than for students who live on-campus.

_____

_____

_____

_____

_____

_____

_____

_____

6. *Hypothesis:* Personal energy use will vary greatly among students due to differences in personal habits, living arrangements, modes of transportation, and diligence in carrying out the exercise.

_____

_____

_____

_____

_____

_____

_____

_____

_____

_____

Lab 10

GLOBAL CLIMATE CHANGE

ACTIVITIES

1. Global Warming—"An Inconvenient Truth"

INTRODUCTION

Our planet is surrounded by a thin layer of mixed gases that we call our atmosphere. This atmosphere provides the residents of this planet with a multitude of vital services. It provides oxidative heterotrophs, such as ourselves, with the supply of oxygen necessary for cellular respiration. It provides photosynthetic autotrophs, which serve as the foundation of most food chains, with a supply of carbon dioxide for photosynthesis and a supply of oxygen for cellular respiration. It plays a role in the water cycle, upon which all life is dependent for meeting water needs. Its presence has helped moderate the temperature of our planet at levels that can sustain life as we know it. The ozone contained in the layer stratosphere helps protect life on the surface from harmful ultraviolet (UV) radiation from the sun.

Perhaps more than any other resource, our atmosphere has fallen victim to the "tragedy of the commons." It is a resource upon which we all depend and to which we have free access but that is not owned or managed by any single entity. As a result, we tend to use such a resource as to maximize our personal benefit or gain with little concern about the long-term effects on the quality or quantity of the resource.

Human actions have degraded our atmosphere in many ways. Increases in the atmosphere in the quantity of suspended particulate matter and in the concentrations of various gases have led to degradation in air quality that ranges from aesthetically unpleasing reductions in visibility to negative health impacts on individual species and ecosystems. Increases in the concentrations of "greenhouse gases" are believed to be the cause of global warming and climate change.

Some of the same gases involved in global warming also contribute to the acidification of precipitation that can lead to a reduction in the pH of surface bodies of water and of the soil, causing damaging effects on plants and animals. Release into the troposphere of "ozone depleting compounds" generated by human activities has led to the depletion of the stratospheric ozone that helps protect the surface from harmful UV radiation from the sun.

Lab 10
ACTIVITY 1
Global Warming—"An Inconvenient Truth"

OBJECTIVES

▪ Understand the concept of the greenhouse effect.
▪ Understand the causes of global warming.
▪ Understand the potential consequences of global warming.
▪ Understand what actions could be taken to mitigate the global warming.

HYPOTHESIS

▪ None.

MATERIALS

▪ "An Inconvenient Truth" DVD

PROCEDURE

1. The class will view the movie "An Inconvenient Truth."
2. Prior to viewing the movie, review the information you are for on the data sheet so that you will know what to be listening for during the movie.
3. Answer the questions that you can during the movie.
4. After the movie, discuss its strengths and weaknesses with your group.
5. Record your response in Tables 10.1, 10.2, and 10.3 on the data sheet.

Lab 10

ACTIVITY 1

Global Warming—"An Inconvenient Truth"

Student Name: _____   Lab Group: _____

TA: _____   Lab Date/Section: _____

DATA

TABLE 10.1 List Five Words/Concepts That Are Important to Understanding Global Warming and Definitions/Explanations for Each

	Word/Concept	Definition/Explanation
a.		
b.		
c.		
d.		
e.		

Lab 10

ACTIVITY 1 Global Warming—"An Inconvenient Truth"

Student Name: _____   Lab Group:_____

TA: _____   Lab Date/Section: _____

DATA

TABLE 10.2. Four *Weaknesses* of the Film's Message about Global Warming, and Supporting Reasons, Identified by Group Members During Post-Film Group-Level Discussion

	Weakness	Reason
a.		
b.		
c.		
d.		

Lab 10
ACTIVITY 1
Global Warming—"An Inconvenient Truth"

Student Name: _____   Lab Group: _____

TA: _____   Lab Date/Section: _____

DATA

TABLE 10.3. Four *Strengths* of the Film's Message about Global Warming, and Supporting Reasons, Identified by Group Members During Post-Film Group-Level Discussion

	Strength	Reason
a.		
b.		
c.		
d.		

Global Warming—"An Inconvenient Truth"

Student Name: _____   Lab Group:_____

TA: _____   Lab Date/Section: _____

DISCUSSION & CONCLUSIONS

For full credit, questions should be answered thoroughly, in complete sentences, and legibly.

1. Go to http://www.epa.gov/climatechange/emissions/ind_calculator.html and use the EPA's Personal Greenhouse Gas Calculator to estimate your total greenhouse gas emissions.

   a. At the end of the "Your Total Estimated Emissions" section, what was the rough estimate provided by the calculator of your household's total emissions?

   _____

   b. *How* did your total emissions compare to the United States average of 20,750 pounds per year for a household of one?

   _____

   _____

   _____

   c. At the end of the "Reduce Emissions" section, by *how* much would you reduce your emissions if you took all the actions listed?

   _____

   d. *What* would your **new** total emissions be?

   _____

   e. *How* does your **new** total emissions compare to the U5 average of 20,750 pounds per year for a household of one?

   _____

   _____

   _____

2. *Did* the movie convince you that global warming is occurring as a result of human activities? Explain *why* or *why not?*

_____

_____

_____

_____

3. If the movie convinced you that global warming is the result of human activities, *did* it motivate you to take personal action? *Why* or *why not?*

_____

_____

_____

_____

4. *What* personal actions do you think you would/could realistically take?

_____

_____

_____

_____

5. *Would* you purchase a hybrid or electric car? *Why* or *why not?*

_____

_____

_____

_____

6. *Are* there energy options for the average citizen that do not contribute to global warming?

_____

_____

_____

_____

7. *Is* global warming an urgent issue in need of remedial action? *Why* or *why not?*

_____

_____

_____

_____

Appendix I

CONSERVATION PROJECT

PRE-LAB READING

Textbook: Chapter 11: Conservation Biology: The Search for Solution.

TIMELINE AND ACTIVITIES

1. **Lab 2**—Each group will be assigned a topic.
2. **Lab 3**—Group members decide who will be responsible for each of the four questions and provide an assignment list to TA. The lists must be signed by all group members.
3. **Labs 4–6**—During these weeks you will need to do the research on the topic and find the best and most appropriate sources (websites and/or papers posted online) that will have all the information to address the four questions. You have to use at least three sources. Make a list of the all sources for the group, format it following the MLA guidelines posted on the BlackBoard, and submit it as a hardcopy to your TA on the **lab 6**. You will not get the points for the Bibliography but the failure to turn it in by the due date or the failure to submit it in a required state/format will result in a **5 penalty points**.
4. **Labs 7–9**—Work with group members on the genre/style of your presentation and identify who will be responsible for each part of the presentation. Make sure each student gets equal amount of information to cover. Each student should make a draft of his/her part before **Lab 10**.
5. **Lab 10**—Work with group members and finalize the presentation during the lab.
6. **Lab 11**—Final group presentation.

Except in extraordinary circumstances, you will receive NO credit for the project if you are not in class the day it is presented.

INSTRUCTIONS

I. PURPOSE

This project offers you an opportunity to investigate a real-world conservation issue of regional interest and to apply the environmental concepts we have discussed in class to the kinds of issues you will be expected to understand as an informed citizen and voter. Working on the project will also allow you to develop and/or improve your skills in research, writing, oral communication, and working in a cooperative, group setting. In addition, it allows you to express your creative side.

II. CONTENT

Your group will be provided with a topic by your TA. Only one group in each lab section may work on each topic. Your job is to research the topic with the goal of presenting the following information, with appropriate attention to detail throughout. Each student in the group will be responsible for researching, answering and presenting the answer(s) to one of the four main questions and providing the appropriate bibliography (see below).

1. **What is the nature and scope of the problem:** What, exactly, is the issue? How widespread is the problem; how long has it been going on? You should include some hard data (numbers, etc.) here and give us a *reason to care about it*. Be realistic and precise—saying that a given problem will "kill us all" or "destroy an entire ecosystem" is almost certainly going to be an exaggeration.

2. **Why it happened:** What was/were the causes? Here, you should explain how and why the problem arose (and/or is currently unfolding) and you should relate the causes to specific ecological principles we've discussed in class. That includes relating specific cause(s) to the general reasons of biodiversity loss.

3. **What are the remedies:** What can be done/is being done to fix the problem? What potential remedies have been suggested? Which remedies (if any) are being implemented? Why? Be sure to address the practical *advantages* and *disadvantages* of competing remedies (remember that no ecological problem is completely one-sided, and most realistic solutions are not going to be simple). Think about who the competing stake-holders might be and what they stand to gain/lose from each potential remedy.

4. **What is the prognosis:** What are the potential long-term consequences of not addressing the problem? This is another area in which you should refer to general ecological principles to help you explain what the small-scale and large-scale consequences of the problem might be. Are current efforts having the desired effects? What needs to be done in the future? What can we do to help? Note that simply stating that "the world as we know it will end" is not a valid prognosis for any of these issues; be more thoughtful and precise than that.

III. MECHANICS

Each student must **use and cite at least 3 references for his/her section**. These can be web-based or print, but be sure they're legitimate. For most topics, you will be able to find good material from websites provided by government, conservation, and/or academic organizations. Be sure to evaluate the source of the information carefully; remember that anyone can put anything on the web, and that conservation organizations vary in their degree of balance and bias. **All references must be cited in the written Bibliography that you submit to your TA on the Lab 6. If you make any changes to the list you will need to resubmit a new bibliography on the day of presentation.** Failure to adhere to these guidelines will result in a reduction in the score.

The group oral presentation should last 15 minutes. Less time means you haven't covered the topic in sufficient depth. All group members must participate and each member should speak for approximately the same amount of time. **Presentations must include illustrations.** When you develop your presentation, be sure you are paraphrasing source material (i.e., putting it into your own words) rather than reading sections of material copied verbatim from your sources. The latter will be considered a violation of the ODU Honor Code and will result in significant grade penalties! **You cannot read your part of the presentation from the cards or from the screen!** This will result in grade penalties. *Presentation must be made on PC not Mac due to compatibility issues.*

Each group can choose the genre/style of the presentation that will be original, creative and informative. See a list of genres on BlackBoard. Your goals are to inform the audience (general public and your peers) about the environmental/conservation issue and to convince people to care about it and act responsible. How you achieve these goals is up to you. Your peers will be judging your creativity and voting for the best group presentation (see the Rubric on BlackBoard). This is your chance to shine!

IV. PRECAUTIONS

Accidents, illness, and other disasters always seem to strike most frequently at the end of the semester. Make sure you share phone numbers, e-mail addresses, and any other contact information you need to take care of last-minute emergencies. And be sure your group plans its work to get things done ahead of time and to ensure that all group members have the resources necessary to give the presentation even if one or more group members are absent. That means being sure that *everyone has a backup copy the presentation AND brings those to class!*

Appendix II

Lab/Activity Title _____

Student Name: _____   Lab Group:_____

TA: _____   Lab Date/Section: _____

Lab/Activity Title _____

Student Name: _____   Lab Group:_____

TA: _____   Lab Date/Section: _____

Lab/Activity Title _____

Student Name: _____   Lab Group:_____

TA: _____   Lab Date/Section: _____

Lab/Activity Title _____

Student Name: _____ Lab Group:_____

TA: _____ Lab Date/Section: _____